COMANCHE MARKER TREES OF TEXAS

COMANCHE MARKER TREES OF TEXAS

Steve Houser

Linda Pelon

Jimmy W. Arterberry

TEXAS A&M UNIVERSITY PRESS　　　College Station

This paper meets the requirements of ANSI/NISO Z39.48–1992 (Permanence of Paper).
Binding materials have been chosen for durability.
Manufactured in China through FCI Print Group

LIBRARY OF CONGRESS CATALOGING-IN-PUBLICATION DATA

Names: Houser, Steve, 1959–, author. | Pelon, Linda, author. | Arterberry,
    Jimmy W., 1961–, author.
Title: Comanche marker trees of Texas / Steve Houser, Linda Pelon, Jimmy W.
    Arterberry.
Description: First edition. | College Station : Texas A&M University Press,
    [2016] | Includes bibliographical references and index.
Identifiers: LCCN 2016010469| ISBN 9781623494483 (flexbound : alk. paper) |
    ISBN 9781623494490 (ebook)
Subjects: LCSH: Historic trees—Texas. | Natural monuments—Texas. | Comanche
    Indians—Texas. | Trees—Texas--Folklore. | Trees—Texas—Identification.
    | Texas—History, Local.
Classification: LCC SD383.3.U6 H68 2016 | DDC 634.909764—dc23 LC record
available at https://lccn.loc.gov/2016010469

DEDICATION

A portion of the proceeds from the sale of this book will be dedicated to a worthy Comanche Nation cause as well as the Texas Historic Tree Coalition in support of their mission and to further research regarding Comanche Marker Trees.

Contents

Foreword

The use of trees to identify a location has always been important in the traditional life of the Comanche. In days of old, our ancestors would mark a tree, or use a tree that was naturally marked and stood out on the landscape, to identify a resource. The tree itself was a resource and could be used to identify various others, such as food, medicine, water, a path, a burial site, or a meeting place. More often than not, these trees marked more than a singular resource.

For nomadic people, it was important to remember locations where activities took place and that, in the oral tradition, held significant information about cultural life. These types of trees are still important and in use today. We utilize the trees within the landscape as a means of following the paths of our ancestors' teachings and connecting ourselves to the country we know as *Numu Soko* (the Comanche word for "Comanche Land"). We still stop to gather medicines and foods, as well as to camp, or, follow various paths that lead us to our destinations, often by the utilization of these trees. Essentially these historic resources are just as significant today as in yesteryear, for we are Comanche—yesterday, today, and tomorrow.

I remember traveling around Comanche country as a child with my grandparents and elders to collect certain resources. There were specific locations where trees offered food or medicine for collection, as well as locations where trees were used for identifying other important occurrences or places. For example, such a tree stood north of our old home place. It was an old, unusual-looking tree that marked the place where my siblings, cousins, friends, and I would gather to play. It was a place where the kids could be found at any time of the day—easily within calling distance in the advent of an important announcement or need. There was no distinction among seasons and its use. It was the kind of tree that stood out in the landscape, calling your attention, begging you to come closer, and directing your activities. Whether it was with or without leaves, we spent hours in and beneath that tree, discovering nature and ourselves. It was a guide of sorts, and a friend as well, who seemed to listen to our stories and protect us from ourselves. And although we children used it as a means of entertainment, adults took advantage of the medicine and materials it offered. To us it was a gentle giant who grew as we did, yet stood still in time. Simply put, it was a marker tree that provided a variety of services.

There was also an area along the creek where certain trees stood tall above the rest and yet seemed hidden in the midst of the grove. Those trees brought great joy to our household when the fruit was added to *tah ah* (the Comanche word for "pounded or jerked meat"). Various nuts and sometimes fruit were added to naturally sweeten the meat as it was pounded or added to other foods. In autumn, these trees would provide fruit for the taking, and usually the family and extended community would gather to collect their offerings. As I reflect upon my early years among my elders, these are only a few of the many examples of support these important trees gave us. What might seem trivial or mundane was, in actuality, a long tradition where identifying, marking, and utilizing a specific tree was important in Comanche life.

So, today we acknowledge that past and look toward the future as we continue to identify, mark, and utilize Comanche Marker Trees.

In 1997, a group of Comanche visited the greater Dallas area to see firsthand a *piapu nakutabai huupi* (Comanche words for "big pecan tree") and met with our good friends Linda Pelon, Steve Houser, and the Dallas Historic Tree Coalition, all of whom were instrumental in recognizing these magnificent trees. From that visit, the first Comanche Marker Tree was formally recognized via a tribal proclamation. As a result of that visit and numerous other meetings over the years to discuss the varying aspects of marker trees and Comanche cultural heritage, as well as share research data and plan trips to see potential trees, it was decided that the time was at hand to produce a book describing this aspect of Comanche culture. What follows is the result of many years and miles of research-gathering among a host of friends with a shared interest in historic preservation, emphasizing Comanche Marker Trees.

JIMMY W. ARTERBERRY
Comanche Nation Tribal Historic Preservation Officer

Preface

I had read about the potential existence of Indian marker trees, and I was intrigued by the subject but completely uneducated about it until, in 1995, I met anthropologist Linda Pelon. At first, I had to admit being concerned about their validity.

In 1996, Linda recognized a tree in East Dallas as a potential Comanche Marker Tree, and she discussed the subject with Comanche officials. In essence, the tree became the first Indian marker tree ever to be formally recognized by proclamation of an American Indian nation. The celebration to recognize the tree changed my life in a number of ways. For an outdoors person, it was moving to hear a Comanche tribal elder explain that his people are "one with nature"; such experiences still give me goose bumps and raise the hair on my arms. It was my connection with nature that had always made me an advocate for trees. But my early bond had always been a gut instinct rather than a clear understanding of the human connection to nature and the interconnectedness of all things. The Comanche officials opened my eyes to many things I had never seen before.

To hold hands with my children, friends, and Comanche officials as we danced and sang to the tree, was a treasured experience. It offered a very different perspective on many things we did not understand but the event touched all those who attended.

We cannot preserve significant trees or cultures that we fail to research, recognize, or fully understand. Indian marker trees are living witnesses to the history of a past civilization and its people's incredible way of life. These trees are a significant part of this nation's cultural heritage and priceless cultural treasures for our current society.

Indian marker trees need to be located and recognized to help ensure that future generations have an opportunity to enjoy them. All trees will be lost over time, so recognizing them beforehand is an important and time-sensitive task. In addition, the elders of the various tribes around the nation who have knowledge of Indian marker trees will not be around forever, and any remaining knowledge they have of these trees must be documented and recorded. The close connection between humans and trees extends throughout time because the two are intertwined. It is hard to discuss the history of one without the other.

This book is written out of a need to educate the public about a subject about which precious little has been documented. It is an introduction to the general topic from a professional perspective. The core subjects are presented in a logical fashion in the first section of the book, with the information on six individual trees officially recognized by the Comanche Nation at the end. For the reader, specific tree descriptions and photos will make more sense after gaining a basic understanding of the subjects covered in the earlier chapters.

The primary objectives in the long term are to find, record, research, and recognize as many Comanche Marker Trees as possible. It is a way to give back a small part of Comanche culture that was once lost.

This book was also developed out of a deep and profound respect for the Comanche way of life and as a way to honor the many significant contributions of that tribe to our nation's history.

STEVE HOUSER

Acknowledgments

We are very appreciative and deeply honored to have the long-term and solid support for our research offered by the Comanche Nation Tribal Elder Council and its tribal chairman, Mr. Wallace Coffey, and former tribal chairman, the late Johnny Wauqua. We are also very grateful for the long-term support and education provided by the Comanche Language and Cultural Preservation Committee. Special thanks to the Comanche Nation Tribal Historic Preservation Office staff, and especially Theodore Villicana.

Our accolades go to the Texas Historic Tree Coalition's board of trustees: President Mary Graves; Treasurer Kirbie Houser; Secretary Sara Beckelman; and board members Bill Seaman, Jim Bagley, and James Alderman. Additionally, our Indian Marker Tree Committee volunteers: Ann Bagley, Anne Weis, Bill Seaman, Bob Richie, James Alderman, Jim Bagley, Kathleen Graham, Kirbie Houser, Mary Graves, Monica Morrison, R. J. Taylor, RuthAnn Jackson, Sara Beckelman, Sharon Bauer, and Thomas Templeton. We also appreciate the work of the dozen newest members of the Indian Marker Tree Committee, whose research will undoubtedly be reflected in the results of our future efforts.

Our appreciation goes to Dr. Daniel J. Gelo at the University of Texas, San Antonio, and his associates in the Department of Anthropology, Thomas Hanson and Jason Roberts, for their efforts to develop various maps of natural resources in association with recognized and potential Comanche Marker Trees.

Our gratitude goes to Dennis Downes, renowned artist and trail marker tree researcher, for his many years of friendship and support, discussions, notes, and photos. We are grateful for the potential Indian marker trees he sends our way, as well as the information he shared with us in his book *Native American Trail Marker Trees: Marking Paths through the Wilderness*.

Our plaudits go to Don Wells for his friendship and support, and for the information he shared with us in his book *Mystery of the Trees: Native American Markers of a Cultural Way of Life That Soon May Be Gone*. Plaudits, too, to the good folks at Mountain Stewards for their efforts to expand public knowledge of this subject through their Web site and newsletters.

We acknowledge our vast indebtedness to Jimmy W. Arterberry, Carolyn Skei, Bill Seaman, RuthAnn Jackson, and Barbara Houser for their hard work in editing.

Our hurrahs go to Bill Seaman for the great photographs he provided and Jim Bagley for two great maps.

Finally, thanks to David Richardson, Brad Ward, Melinda Adams, Anne Weis, Kevin and Roseanna Webb, George Blackmon, Ada Lester, Karl Williamson, Tai Kreidler, Holle Humphries, Michael Jones, Keith and Irene Henry, Richard "Rich" Denny, LeeAnn Weaver, and Ted Braun, among others, who took time out from their lives to pass along photos, information—and oftentimes both—of potential Comanche Marker Trees. Over the years, more than 450 people have submitted trees with potential.

STEVE HOUSER

Introduction

LINDA PELON

When I began research for my thesis, *Issues in Penatuhkah Comanche Ethnohistory* (Pelon 1993), Comanche land use in Texas had been identified as an important issue. The Penatuhkah Comanche band had controlled most of Central Texas and had been prominent in peace negotiations with both the Republic of Texas and, then later, with the US government as Texas was annexed into the United States in the late 1840s. The "Indians" who fought the early and legendary Texas Rangers led by Jack Hayes were Penatuhkah chiefs and warriors. They successfully defended their immense homeland from an unending tide of European and American invaders.

The leaders of the invaders who encountered them recorded much information about this courageous and intelligent Comanche band. Organizing and documenting this information had tremendous potential to reconstruct the history of this southernmost group of Plains Indians who were on the front lines of protecting an American Indian transportation corridor across the Great Plains. This critical transportation corridor connected allied Indian groups from Alberta, Canada, throughout Texas, and into Mexico and the American Southwest. How did they navigate and use this immense and varied landscape? How did they protect it? How did they communicate across it? Some very intriguing information and possibilities began to emerge as answers to these questions. Smoke signaling, rock art, and marker trees became important issues for answering some of these questions. This book is focused on Comanche Marker Trees as landscape navigation aids and organizes information from two decades of research from multiple perspectives.

While the focus of this research has been on Comanche Marker Trees, other information emerged, indicating that many types of trees were an intimate part of Comanche ways of life. The tree taxonomy created by Comanche Nation Tribal Historic Preservation Officer and research partner Jimmy W. Arterberry provides a glimpse into this relatively unexplored research topic. The limited information provided raises the question, "Were trees as important to these Plains Indians as buffalo?" Additional research is needed to address this and other questions. At the very least, it is obvious that those who want to understand Comanche land-use patterns in Texas need to include tree use in their research.

Information was gathered about possible Indian marker trees starting in that first year of research into Texas Comanche land use. Ranchers and descendants of Texas pioneers shared information about unnaturally shaped old hardwood trees on their properties. Some of these trees were at strategic locations, such as low-water river crossings, or on archeological sites with evidence of former Indian campgrounds. Other informants shared oral historical information passed down their family lines about the significance of these trees to Indians (marking gathering places for treaty talks, known Indian trails, and more). These individuals seemed convinced that these trees were intentionally bent to mark special places. Most archeologists I spoke with about these trees stated with absolute authority that there were no such things as Indian marker trees. They also admitted they had heard of them but had never done any research on this topic. One archeologist justified his denial of these trees by saying that a rancher on a site he excavated told him a "story" about an odd-shaped tree on that property. The archeologist stated that he dismissed the story because the site was a prehistoric site and the tree could not possibly be that old. Such a position was and is intellectually troubling. Most significantly, I wondered how these scientists could express these opinions with such authority and certainty when they admitted that they had never researched the topic. It was also troubling that an unusually bent tree was dismissed on an archeological site with an associated ranch history that distinguished it as an Indian marker tree. I have visited sites that are historically documented Comanche campsites that no longer had evidence of a historic Indian occupation but still contained artifacts and features associated with prehistoric use. My conclusion was that the very thin and fragile top layer of historical occupation had not survived the more than 150 years of surface collection of Indian artifacts, floods, erosion, and other human and natural forces that remove the top layer first.

One site that fits this description is near Santa Anna, Texas (named after the Penatuhkah Comanche war chief Santanna/Santa Anna, whose stronghold was located there in the 1840s). This site was visited in the early 1990s by Comanche leaders who first asked, "It is obvious that there were Indians here, but how do we know there were Comanche people here?" They answered their own question when they discovered a stone feature they recognized as a Comanche ceremonial site. Within months, the Comanche Nation officially claimed this site as a Traditional Cultural Property, and the property owner was honored as a "Land Keeper" for the Comanche Nation.

In summary, those who have done no research on Indian marker trees—even if they are scientists/archeologists—are not qualified to determine if Indian marker trees exist. The surviving archeological evidence at sites can be misleading. Some sites with no evidence of a Comanche occupation indeed had one, but that layer of occupation no longer exists. Records left by military leaders, scouts, and Indian agents can be used as evidence to document many of the sites significant to Comanche land use. Additionally, contemporary Comanche people are heirs to traditional information not known to archeologists that can tell us much about these sites and their uses.

Our research goal for the Comanche Marker Trees in Texas project was to collect information on nominated trees either to prove they are marker trees or to rule them out as candidates. Concerning the shape of these trees, the hypothesis was that a *half-moon* or *rainbow* shaped trunk was used as a Comanche style for their marker trees. This shape had already been documented as that of a Comanche Marker Tree by the Texas Forest Service in its book, *Famous Trees of Texas*. In the early stages of our research, we discovered a different shape for a marker tree. This shape had a trunk bent to run parallel with the ground and then bent again at a right angle to allow the tree to grow skyward once more. These trees, called *thong trees*, were reported to be Cherokee Marker Trees. This information created awareness that at least one other style of marker tree existed and was associated with Cherokee culture. Since there was a past Cherokee occupation in Texas, our team suspected that we might encounter some trees in this style. And there were several. We noted the locations and other available information but did no additional work since research for this project was focused on Comanche Marker Trees.

Recently, other unnaturally bent trees have been discovered in association with what may be Comanche Traditional Cultural Properties. Research continues, and there is a growing body of evidence that this simple and intelligent strategy for creating signage was widely used by this nomadic culture. This book is focused on the Comanche Marker Trees in the half-moon or rainbow shape. Some of these trees may have occurred naturally. The live oak, documented by the Texas Forest Service, may be one of these. Perhaps meeting at a naturally bent tree gave these nomads the idea of bending other trees to mark meeting places. However, many trees found bent in this shape (bur oaks, pecans, red cedars) do not bend naturally. Some of these trees have scarring consistent with having been tied down, and they occur in association with intersections of Indian trails, significant campsites, and important natural features. In two cases, there are bent trees on both sides of low-water crossings for major Indian trails. It is highly unlikely that nature provided bent trees in all these culturally significant places.

Over the first decade of our research, several trees with the rainbow/half-moon shape thought to be a Comanche style of marker tree were nominated. It was easy to rule out most due to insufficient age. The sites of others had no evidence of a former Comanche occupation and no significant natural features in association with the trees. It was and is frustrating that a growing number of these trees cannot be either ruled out or proved to be marker trees using mainstream standards of proof. And this is the point where two competing and often contentious anthropological perspectives become relevant. Scientific anthropologists are most concerned with using the scientific method to find "proof." Humanistic anthropologists are highly motivated to "increase understanding." These two approaches have the potential to be complementary but often are competing and conflicting instead.

From the perspective of humanistic anthropology, these trees have provided valuable information. Working with the hypothesis that bent trees that cannot be ruled out may have marked significant features, or provided important information to Comanche travelers proved to be very productive. When combined with the recently emerging landscape model for recognizing and understanding archeological sites and Traditional Cultural Properties, a strong case can be made for the significance of these trees to landscape navigation. The section "Applying the Landscape Model" in chapter 3 will explain this approach in its discussion of information relevant to understanding Traditional Cultural Properties that include environmental features with no known archeological evidence of occupation.

The chapters documenting individual Comanche Marker Trees will use the landscape model to provide supporting evidence for recognizing these trees as Traditional Cultural Properties. Our position has always been that it is the Comanche Nation who should decide whether or not these trees are Comanche Traditional Cultural Properties if they occur in areas documented as part of the former Comanche homeland. As a sovereign nation they have both the right and the expertise to evaluate these sites according to their own cultural traditions.

PART ONE

The Process

Basic Information Regarding Indian Marker Trees

CHAPTER 1

STEVE HOUSER

Introduction to Indian Marker Trees

L iving in harmony with nature has been a way of life for American Indians. They relied on nature for all their needs. Traveling from place to place required good navigational skills, and directions provided along the way were helpful. Because there was a need to develop a method to mark common trails, many American Indian tribes used trees not only to mark a trail but also to signal the presence of important features, some of which were critical for survival.

Trees were used to guide American Indians to sources of water, good places to cross rivers, campsites, and other important natural features. To those who could interpret their meaning, trees were often similar to present-day lifesaving road signs.

An Indian marker tree is one that marked important resources, an event, or a location. One form of marker tree was the bent tree, which is also called a *turning*, *pointing*, or *leaning tree*. These trees were often tied down with a yucca rope or a thong of animal hide, which is where one of the common names, *thong tree*, originated. In some cases, one tree was tied to another or rocks were also used to turn tree growth in the desired direction or coax it into the desired shape.

Comanche Nation Tribal Historic Preservation Officer Jimmy W. Arterberry once noted that the Comanche would be more likely to use a yucca rope to tie a tree, due to the fact that animal hide was a valuable resource used for worthier purposes. He also noted that yucca cordage has been found at numerous archeological sites and it has been used to make sandals, baskets, mats, and ceremonial regalia, among other uses. It will last a very long time under the right conditions. If the grasses were as thick as reported in historical times, then there is the probability that such cordage entwined with additional grass support would last even longer. The variables of its longevity are significant in relation to the species, location and general environment of the time. The strength and longevity of yucca rope would make it ideal to hold a tree for a number of years. In addition, yucca plants were likely to be easier to collect than animal hides.

Today some people call these "old road signs" *Indian marker trees.* Others refer to them as *trail trees, thong trees,* or *culturally modified trees.* The Comanche people typically use the general term Indian marker trees for all trees around the United States and Comanche Marker Trees for trees related specifically to Comanche culture and heritage. Chapter 4, "Comanche Marker Tree Taxonomy," explains in further detail how native tree species were used for very specific purposes.

What Makes a Comanche Marker Tree?

The presence of American Indian tribes in Texas predates the state itself. Spanish and French records document tribal peoples in Texas upon explorers' first excursions into the region. Some of the tribes have been here at least 13,000 years, perhaps longer. The Comanche were the relative newcomers in the group.

Tribes with primary dominance included the Caddo Indians in East Texas, the Tonkawa Indians along the coast, and Lipan Apaches in far South Texas. The Comanche claimed Central and West Texas, as well as North Central Texas. The Penatuhkah were the largest and best-known band of Comanche to inhabit Texas, and they were located primarily in the central part of the state. The Waco (Huaco), Wichita, Keechi, and Tawakoni tribes were present in the North Texas area. The Cherokee were one of the most recent tribal groups to enter into Texas, doing so as a result of the Indian Removal Act of 1830.

In the long term, not all these tribes survived and held together as groups over the years. The Waco (Huaco), Wichita, Keechi, and Tawakoni tribes have banded together because they were so few in number after being forced to move to Oklahoma and stricken with disease. Discussions with officials from that group revealed that they were unaware of their ancestors bending trees for a specific purpose. Likewise, other tribes known to inhabit Texas have yet to formally recognize any unusual trees that may be associated with their tribe. A number of these tribes are known to have existed in Texas for a long period of time, but the timeline for their occupation of specific areas is not well recorded. The Comanche, on the other hand, have particular timelines associated with their activities over a long period of time. We hope that other tribes will find information that was previously lost regarding their use of trees as markers. To date, the Comanche are the only tribe to inhabit Texas that formally recognizes marker trees as an important aspect of their culture.

Significant Trees or Groves as Markers

Chapter 7 discusses the fact that forces of nature can create a bent or odd-shaped tree. It is likely that many were used to aid in navigation or path-finding in the past, even if humans did not create them.

It is important to recognize that unusually shaped trees, groves of trees, or large trees that stand out would also have been types of Comanche Marker Trees as well as the markers for early settlers and land surveyors. In the past, directions often included a large tree or a grove of trees as a turning point or a marking point to aid in navigation. Going the right direction was important, and with few if any maps, directions were provided by geological features and trees. "Go up the creek until you hit the sharp bend, and head east until you find the big tree out in an open field, then head north." Many types, sizes, and even groves of trees can be a form of marker. Early property surveyors often noted a large tree as a marking point on their surveys.

A large tree would not point in a particular direction, of course, whereas a bent tree could be more precise in indicating a good general direction. This is why trail marker trees, also known as turning trees, were created.

Not All Comanche Marker Trees Were Bent

As Arterberry notes (a bit ambiguously) in chapter 4, "All trees can be marker trees, but are not." The fact that a tree has an odd or bent shape does not make it an Indian marker tree. By the same token, the fact that a tree has a normal shape does not lead to the conclusion that it is *not* an

Indian marker tree. A number of normal-shaped trees, such as the Treaty Oak in Austin, have a customary shape but are significant to American Indians. Arterberry also states in chapter 4, "Treaty/council trees were used as council places and to make a treaty, but they are not bent trees." Often, even a normal-shaped tree provided a valuable service to the Comanche.

A Comparison of Indian Marker Tree Uses

Recorded history and discussions with American Indian officials uncover many of the past uses for trees inside and outside the state of Texas. Some of the uses are consistent around the nation, such as a trail marker tree or a treaty tree. Marking trails appears to be the most common use of these trees, and they are found along trails in many parts of the US. They are known to be part of a very old trail system used by various tribes over the years. Trails were often created by animals and later traveled by humans. Some tribes, such as the Comanche, were known to be nomadic buffalo hunters and often moved from place to place using existing trails. Since the Comanche did not enter Texas until the early 1700s, some of the trails may have already been in existence and unfamiliar to them. Battles between various Indian tribes, as well as battles with early settlers, would have caused tribes to travel, sometimes unexpectedly. Travel also included hunting and gathering trips in search of food. The more a tribe had to travel into unknown territory, the greater the need for trail markers.

Treaty trees witnessed the signing of an important treaty, and tribes around the country recognize such trees. A treaty tree is both a form of Indian marker tree and a historic tree. The definition of a historic tree can vary among different groups. Most would agree that it is a tree that shares a significant event within a particular place in time. The famous Treaty Oak in Austin, Texas, is a good example. This live oak, *Quercus fusiformis*, is the last survivor from a group of fourteen oaks that once served as a sacred meeting place for the Comanche and Tonkawa tribes. Stephen F. Austin is reported to have met with American Indian tribes in that group of oaks to negotiate and sign the first Texas boundary treaty.

Large and old trees were consistently used as markers in many parts of the US due to their prominence (or visibility) in the landscape and their ability to pinpoint a location. The same is true for groves of trees comprised of large numbers of the same tree species. In any part of the country, a large old tree would stand out, regardless of its location, and likewise, a grove of trees comprised predominantly of one tree species could have been an Indian marker tree grove.

An Indian marker tree may also have been used to mark a boundary for a particular territory. Such trees would act as a sign that others could recognize. Research continues regarding the potential for this type of Indian marker tree to be found in Texas. It is possible that this use may be more consistent among tribes in various parts of the country than previously thought.

Other uses of trees are more individual to the tribes they represent, such as the Ute Indian Prayer Trees in Colorado or the Comanche Storytelling Place Tree in Dallas, Texas. The Ute Indians are the only tribe, to date, that recognizes a prayer tree as an important part of their culture. A pine tree was tied down to serve as an altar on which prayers would be held that would go ultimately to the Creator. The Comanche are the only tribe to recognize a storytelling place tree. It is not clear when or where the bending of trees for a specific purpose started. However, the consistency of this practice among many tribes around the country would indicate that it was a long-standing cultural tradition.

The consistency of use among tribes is noted by Arterberry:

> Prior to entering into Texas, the Comanche had migrated out of the Great Basin area, then southerly into the Ute territories and became allied with their linguistic relatives. They spent many years living amongst the Ute before entering into the Southern Plains of Texas. The Comanche brought with them a technique of tree usage that served a host of purposes. Not only were these trees used to mark trails, springs, meeting locations, plant and geologic resources, they identified boundaries, events and other activities that were important to Comanche cultural history. The Comanche were keenly aware of the landscape and all of its offerings, which included a variety of trees. It is geography that dictates tree usage.

Why Most Are Not Aware of These Trees

Many years ago, American Indians became reluctant to share details regarding their customs with outsiders. They had lost the basic necessities for their way of life. The Comanche's painted trees and carved trees, as well as other trees obviously related to the American Indian way of life, were often removed by early settlers. Bent and less-obvious Indian marker trees often survived because only the American Indians knew of their significance, and tribe members were instructed not to talk about them. As a result, it is difficult to uncover the trees' secrets from a past that was buried long ago. Add to that the fact that Traditional Cultural Properties and Indian marker trees are often in the way of future development or redevelopment. As a result, society often greatly underestimates or ignores the true value of these cultural treasures. The American Indians have always seen the need to live within the balance of nature, and they have a great reverence for all the glorious things that nature provides. Many tribes integrated nature with agricultural principals to help improve their own natural food sources. They are the ultimate stewards of our natural world, while society in general does not appear to hold these same values.

The Dallas Historic Tree Coalition

Efforts toward researching Indian marker trees were established by the Dallas Historic Tree Coalition. The coalition was founded in 1995 as a nonprofit 501(c)(3) organization comprised of volunteer advocates for irreplaceable trees in the Dallas-Fort Worth region. Irreplaceable trees include historic, heritage, champion, Indian marker, and other trees deemed significant.

The group was originally formed to encourage the preservation of eighty large shade trees at a hospital in Dallas. The coalition raised funds and ran a double full-page announcement in the *Dallas Morning News* on Sunday, July 23, 1995. It announced the formation of the group and its mission and goals, but it also covered the many benefits of trees, the growing urban heat island problem, challenges that tree removal poses to quality of life as well as noting that the land being built upon is an essential part of Texas history. Furthermore, the announcement discussed the trees and offered a proposal from the coalition that found common ground with hospital developers and helped meet the goals of both groups. The effort successfully preserved around seventy percent of the trees, including two very large bur oaks, *Quercus macrocarpa*. It became clear that a group of dedicated individuals can make a difference, which provided a great deal of inspiration toward expanding the group's efforts.

The first president of the Dallas Historic Tree Coalition was Dr. Mary Ellen Bluntzer. Founding members included environmental icons Ned and Genie Fritz and Janice Bezanson, among others. Early coalition supporters included anthropologist Linda Pelon, who became actively involved in the group and served as the executive director for a number of years. The coalition was aware of a book produced by the Texas Forest Service titled *Famous Trees of Texas*, which included a tree that was "reported" to be an Indian marker tree. Linda Pelon noticed a potential Indian marker tree in her neighborhood and brought it to the coalition's attention. The subsequent work by the coalition to officially recognize the first Comanche Marker Tree was essentially the beginning of a long-term effort to formally recognize more of these trees in the future.

Over time, committees of volunteers were established to find, research, recognize, and celebrate Indian marker trees, as well as historic trees. These included the Indian Marker Tree Committee, the Historic Tree Committee, and a Historic Research Committee. The Historic Research Committee was established to delve into the history of a location or tree while it was under review by the other two committees. Historic research of many types can provide facts and information to support a historic tree nomination or the submission of a potential Indian marker tree.

Adoption of a New Name and Expanded Mission

Following many years of research into Indian marker trees in Texas and beyond, the Dallas Historic Tree Coalition's board of trustees made the decision to establish and operate as the Texas Historic Tree Coalition. The board's decision is recognizant of the need for the organization to formally recognize significant trees around the state. The long-term goal is to encourage other groups of historians, arborists, American Indians, anthropologists, and like groups to form their own historic tree coalitions in their particular regions.

The Texas Historic Tree Coalition works diligently to uncover information about potential Indian marker trees and will continue to develop a public understanding of these trees and aid in their proper recognition. The Indian Marker Tree Committee currently receives an average of one new potential Indian marker tree submittal every two weeks, each requiring further investigation. Although more than 450 trees have been submitted, the committee keeps files on approximately 150 trees that are currently under investigation. As an all-volunteer organization, members find the tasks included in these investigations to be both enjoyable and exciting, though at times overwhelming.

Why Recognition Is Important

The purpose of discussing Indian marker trees in a public forum is to increase awareness of their existence, to recognize them as living witnesses to our history, and to preserve them through documentation, should they be lost in the future. The trees provide lessons about our past and lessons yet to be learned, but their life expectancy is limited. In the past, drought, old age, fire, land clearing, development, and other factors caused the loss of a number of trees under investigation. This underscores the urgency of documenting as much information as possible about the trees that remain.

Six Comanche Marker Trees Officially Recognized

Photos of Indian marker trees, both recognized and under investigation, can be found on the Texas Historic Tree Coalition Web site, as well as in online listings provided by a consortium known as the Texas Tree Trails, which also records historic, champion, and other significant trees. We offer here a glimpse of the six trees formally recognized by the Comanche Nation.

The chapters to follow explain much more about subjects related to Indian marker trees and specifically these Comanche Marker Trees.

California Crossing Comanche Marker Tree.

Gateway Park Comanche Marker Tree. Photo courtesy of **Dallas Morning News** *photographer Summer Zak*

Cedar Ridge Comanche Marker Tree.

Irving Escarpment Comanche Marker Tree.

Bird's Fort Trail Comanche Marker Tree.

Storytelling Place Comanche Marker Tree.

Early Recorded History of Indian Marker Trees

STEVE HOUSER

CHAPTER 2

In the book, *A Texas Legacy, The Old San Antonio Road and Camino Reales, A Tri-Centennial History*, A. Joachim McGraw and Kay Hindes attributed trees with a cross carved into their trunks to early Spanish colonists. They write, "Field inspections showed a distinctive cross blazed into the cambium of a Post Oak adjacent to a narrow, deeply rutted area. It is believed the proximity of such a blazed-marked tree along the estimated route of the Old Laredo Trail may have been directly related to the actual colonial route" (1998, 161).

A second tree was also noted. Although these trees were not Indian marker trees, they were a form of trail tree marker created by the early Spanish settlers in the late 1700s. They could be considered "Spanish marker trees." And if you take into account that various trees were often used as boundary markers by early surveyors, those trees could also be considered "American marker trees."

There is a variety of books, old photos, news articles, and maps, as well as recorded history, which indicate the presence of Comanche or other tribes in various parts of Texas. A few worth mentioning include the maps found in *Comanche Land* by J. Emmor Harston or *Texas Indian Trails* by Daniel J. Gelo and Wayne J. Pate. Similarly, articles such as "'Comanche Land and Ever Has Been': A Native Geography of the Nineteenth-Century Comancheria" by Daniel J. Gelo also provide resources for research on the subject.

These types of records provide critical information that helps us find a direct connection between a site, including its tree(s), and the presence of the Comanche in the past. The Indian Marker Tree Committee searches diligently for any early records directly related to the Comanche and these types of trees. Anyone with information related to Comanche history of a site is encouraged to contact the Texas Historic Tree Coalition.

In the article "'Comanche Land and Ever Has Been,'" Texas Ranger Rip Ford notes that the Comanche used elevated points for reconnaissance as well as for announcing returning war parties. Elevated points were important and a primary reason we now research topography in an area. Tally marks were left on the ground of the high point or in the nearby woods. A small branch or other object was stuck in the ground, leaning in the direction the other Comanche were moving. The article notes that saplings were tied down to mark a trail to water, or the Comanche might split the trunk to mark a water hole. The article also records three other trees thought to be Comanche Marker Trees, two in Burnet County and one at San Pedro Springs in San Antonio (2000, 306).

A review of official records shows, for example, a survey performed by Warren Ferris in the Dallas area between 1839 and 1842. In addition, there is evidence that J. J. Beeman was granted 640 acres in the North Dallas area in 1842. Properties can be tracked through such reviews. These are the kinds of historical records we prefer to find on all properties we investigate because we can follow ownership from the time the property was claimed up to the present day. Of course, early or present survey records may not mention a large or bent tree on a site, although we always hope to find one that was officially recorded many years ago. *(Note: More information on Dallas area history can be found under individual tree profiles.)*

Recorded history shows that the Comanche were known to paint or leave a carving on a tree—a form of pictograph. Although pictographs are typically thought of as paintings on rocks, painted sticks or pieces of bark as well as paintings on the trunks of trees are also legitimate pictographs.

Trees with such pictographs are recorded in *Picture-Writing of Texas Indians*, by Alvin T. Jackson, field archaeologist. Jackson notes, "There seems to be little doubt that some genuine Indian picture-writings were placed on trees in certain parts of the state" (1938, 319). Reports at the time noted that the bark was often stripped off but could be left in place and painted with red or blue paint. The designs were similar to those found in rock shelters in the region and included "all sorts of zigzag lines and curliques" (320). These designs were located three to five feet from the ground on some trees. An Indian head wearing a feather headdress in red and trimmed in black and yellow was reported to have been found on a white oak.

Jackson notes an account of paintings on a group of large cedar trees located in Palo Pinto County near Turkey Creek and known as the "Painted

Camp." Hundreds of paintings on cedar trees were reported, and another group of similar cedars may have existed, as well.

The March 1996 article, "Searching for Turning Trees and Other Special Trees of the Comanche" by Anna Jean Taylor and Comanche Nation Tribal Elder Melvin Kerchee Sr. outlined a great deal of information. The article notes the potential locations of numerous significant trees that date to the early eighteenth century.

> Kerchee learned of turning trees from his grandfather and other elders who spoke about the times before 1875, when the last of the Comanche surrendered to the US military. During those earlier days were special meeting places along the plains' edge, distinguished by good flowing springs and large or unusual-looking trees known as Turning Trees. . . .
>
> Unfortunately, the exact location of many of these places was not passed on. Kerchee observed that the Comanche custom of pointing things out with one's lips (pointing with fingers is considered rude) added to this loss of knowledge. During later day auto drives through the Texas panhandle-plains, Kerchee's grandfather showed where certain Turning Trees were, pointing with his lips, leaving Kerchee with a somewhat vague impression of the place intended. . . .
>
> During 1973 or 1974, Perry Bowser of Plainview, whose family settled an area south of the present-day town of Lubbock, told Archaeologist Joe Hays of a group of special trees. Bowser saw these trees before they were cut down. According to family lore, the Comanche had dug up twelve saplings, split and inserted stones into the base of each, and planted them around a good spring; the stones dwarfed the trees so they would not cover the spring. This site was located along the Comanche trail to Mexico. (1996, 1–2)

The article notes several large cottonwood trees in the Panhandle area and Southern Plains that were previously recognized. As a side note, it is not clear which species of cottonwood is involved. Benny J. Simpson's book *A Field Guide to Texas Trees* notes, "The High Plains also has only one species, *P. deltoides* var. *occidentalis* (Plains Cottonwood)" (1999, 237). Stan Tekiela's *Trees of Texas Field Guide* states, "Several subspecies of Eastern Cottonwood are found in Texas. The Plains Cottonwood (*P.d. occidentalis*), essentially a smaller version of Eastern Cottonwood, occurs in the panhandle and can tolerate drier soils" (2009, 195).

Taylor and Kerchee also noted trees in other parts of Texas.

Beyond the southern plains, in the Hill County of Central Texas, also were trees special to the Comanche. According to Bruseth and Kenmotsu, a descendant of early Euro-American settlers recalled hearing his grandfather speak of a misshapen tree on the family's property near the San Antonio-to-Blanco road. This tree was said to have been used as a trail marker by the Indians. Reportedly, similar claims are made in the diaries of two other early Central Texas settlers.

In Wood County, early surveyor G. W. Cowan saw several trees with carved designs reportedly made by Comanches, located along the Comanche Trail. (Wood County is in East Texas).

Trees painted by Indians also are noted for Edwards, Palo Pinto and San Saba Counties of Texas. Decorations carved onto trees by Indians are reported for Texas counties of Wood, Smith and Van Zandt. (1996, 3)

It is important to note that research was done to find anyone of knowledge regarding the painted or carved trees in East Texas. Most claimed that these trees were removed long ago and no longer exist. It was common in many parts of Texas to hear the comment that trees that were easily recognized as being related to American Indians were removed because they were related to the Indians. Not that long ago, there was a great deal of animosity toward any Indians because someone had lost relatives or friends in past battles. The same is true for the Indians because of the atrocities they suffered. American Indians were ordered not to speak about their culture in the 1800s and early 1900s. Maps and printed materials were taken away and burned.

In May 1941, an article by Lula Kirkland was published in the Dallas Archaeological Society's *The Record*. It notes a series of five Indian trail marker trees in a line along the Irving Trail. All of them are noted as post oaks (*Quercus stellata*), a Texas native and long-living species.

Recently while exploring for signs of an Indian campsite on our Dallas survey, we discovered not only a campsite but near its edge, the rarest of finds in this part of the country—an Indian trail marker. We had never seen one but from a faint memory of a picture of one we recognized this one immediately. What a thrill of joy accompanied the exclamation—oh, an Indian trail marker! It was a pleasure indeed to find this gnarled old tree, mute evidence of the Indians' handiwork still flourishing in this semi-prairie country after almost a century of farm land improvements and occupation by the white man.

Indian trail markers have been found in various places in the eastern and southern United States which includes Texas; but the one here reported on is the first to be found and recognized in Dallas County. Bent trees as trail markers were evidently a universal sign to all Indians. (1941, 44)

Ms. Kirkland describes how the trees were formed. She also notes that if a small sapling was not available, a limb was sometimes bent.

In September 1941, *The Record* corrects the previous article and adds a drawing.

A newspaper article about the society's initial discovery of Indian trail markers in Dallas County brought several reports of other markers. However, only one series of trees proved to be genuine. These are about two miles southeast of Irving, and were reported to the society by Mrs. Robt. S. McKee of that city.

There are five markers in line within a half a mile on the Irving Trail and one single marker a quarter of a mile west of the main trail. All are post oaks. Three of the markers are spaced only 15 feet apart, pointing in the same direction. (See the illustration, traced from a photograph.) One of the markers is a rare case where a limb was broken down to mark the trail.

For additional information on Indian trail markers, see articles by Dr. Raymond Janssen [sic] in *Nature Magazine*, Aug–Sept 1938; *Natural History*, Feb 1940; *Scientific Monthly*, July 1941.

Nature can create bent trees, but it is not common to find three in a row with this type of spacing and with all three pointing the *same* direction. It would be difficult for natural forces to have created these trees. A large tree can fall on small trees or saplings growing underneath, but for two or three trees to be bent in the exact same shape and direction, and for them to hold their shape over many years, would require very specific circumstances. As far as natural forces are concerned, wind could not conceivably create these trees, leaving only snow or ice buildup as potentially responsible for their shape. The odds of such forces bending three trees in the same direction may be plausible, but unlikely.

Although research did not find any record of a bent tree on the site, Turkey Knob is a high point in the Dallas area and a landmark known to be an Indian campsite and a favorite hunting ground. It is also known to be connected to a trail from California Crossing, a low-water crossing point

Three trees spaced 15 feet apart along the Irving Trail.

for the Trinity River, which is one mile to the north. *(Note: More on these sites to follow in part 3, "Profiles of Recognized Comanche Marker Trees.")* The story of these sites was covered in three previous articles (1955, 1956, and 1965) in the *Dallas Times Herald* regarding the purchase of Turkey Knob by the University of Dallas. There was also a 1965 article in the *Dallas Times Herald* regarding Turkey Knob and the California Crossing. These articles have been important in establishing the location of early trails and an important Indian hunting ground.

Turkey Knob provided the Indians with all the advantages of a high point, along with a hunting ground and a nearby river. The *Dallas Times Herald* articles state the site was a landmark once used by wagon trains heading west across the nearby California Crossing of the Trinity River. The site played a vital role in the great movement westward, according to Darwin Payne, *Dallas Times Herald* staff writer. On March 22, 1965, the Dallas Times Herald ran an article by Payne which noted, "The river crossing today is remembered in a road there that is still called California Crossing. A park marks the actual spot of the crossing, although a dam has changed the river's nature." These types of articles help in piecing together parts of history that were often not noticed or recorded. Other early recordings of important sites or trees may yet be discovered.

Reviewing past observations of existing Indian marker trees in Texas provides a closer look at the subject from a historical perspective and helps us better understand the trees we are finding today. The past provides clues that help us uncover the mysteries behind some of the existing trees in our files and those we may find in the future.

A National Perspective

The connection between humans and trees goes back to the beginning of time. The human manipulation of trees has occurred throughout the ages and upon many different continents. In British tradition, trees were often planted and manipulated to form a hedge through processes known as plashing, pleaching, or hedgelaying. Top growth was cut back to cause a thickening of the trunks, and lateral limbs were woven together to form a natural fence or hedgerow. This process was also practiced by early settlers in this country.

In an article titled "Lopped Trees of Long Island," Philip Marshall of Yale University notes, "One of the curious features of the landscape of Long Island is what is termed 'lopped trees,' old oaks (*Quercus* spp., usually *Q. alba*) strangely deformed with right-angle bends in their boles near the base, found along property boundaries (often on roadsides) and the borders of farmland, and sometimes in the middle of the woods" (2001, 37). The article notes the uncanny resemblance to "trail marker trees" found across the Midwest and includes photos from the geologist and researcher Dr. Raymond E. Janssen.

The photos show examples of trees previously researched by Janssen. The Marshall article also notes that the Lopped Trees of Long Island were never confirmed to be marker or trail trees. Marshall points out that the pruning or manipulation of fruit-bearing trees to force more fruit growth is another example of human influence on trees that has a long history.

A national perspective on Indian marker trees is provided in books such as *Native American Trail Marker Trees: Marking Paths through the Wilderness* by Dennis Downes with Neal Samors and a foreword by Lamar Marshall; and *Mystery of the Trees: Native American Makers of a Cultural Way of Life*

Illustrations of typical Trail Marker Sign photos by Raymond E. Janssen. The tree on the left is an example of how sometimes trees took root where the tip touches the ground.

That Soon May Be Gone by Don and Diane Wells with Dr. John Nardo, Robert Wells, and Lamar Marshall. Both books were published in 2011 and share extensive information on some of the earliest recordings of Indian marker trees. Another book that provides basic information is *Indian Trail Trees* by Elaine Blohm Jordan, published in 1997.

Early recordings of Indian marker trees go back as far as the mid-1800s. According to Downes, "Some of the most well-known advocates of the Trail Marker Trees in this area were: the Chicago Academy of Sciences, landscape architects, local historians and historical societies, the National Society of the Daughters of the American Revolution (DAR), Boy Scouts of America, Girl Scouts of America, Garden Clubs, Arboretums, and forestry departments. One of the earliest and most in-depth photographic studies was completed by Bess Bower-Dunn, a Waukegan historian; her study began in 1912 (2011, 45). Downes also credits the DAR for some of the earliest photographic documentation of the trees, placing bronze tablets near them explaining their significance and having special ceremonies to commemorate them.

Downes also notes early recordings of these trees: "The Lake County Book of Records at the Recorder's Office holds a survey from 1852 that has a sketch of a Trail Marker Tree completed by an official Lake County Surveyor, Mr. Hale. In 1912, John Halsey, LL.D., of Lake Forest College, talks about two notable Trail Marker Trees in Lake County in his book *A History of Lake County Illinois*" (2011, 46).

Early recordings were also noted by Don Wells in the *Mystery of the Trees*: "Frank Reed Grover (1858–1919) was one of the first to write extensively about Indian trails and Indian trees. . . . Beginning in early 1900 and extending to his death in 1919, Mr. Grover did research and documented the Indian history of Illinois in major historical writings" (2011, 8).

Mr. Grover's articles were recorded in major Illinois historical writings, with one of the first reports being published in 1901. As a result of his articles, interest blossomed around Chicago, and details of a number of trees and accompanying photos were printed in the *Wilmette Life Newspaper* on March 23, 1939. The same paper reported on a trail tree and showed photographs of Chief Evergreen, a Potawatomi Indian, and a friend named Raymond Gloede, in September 1965. Mr. Gloede was a photographer who wrote a letter to General Able Davis, encouraging him to recognize, preserve, and maintain "old Indian trail trees."

Katherine Stanley Nicholson wrote a book titled *Historic American Trees*, published in 1922, which further discussed trail trees and noted a "well known line of 11 White Oak Trail Marker Trees that stood in perfect alignment and led from Highland Park [Michigan] around the marshes to another Indian village" (qtd. in Downes 2011, 50).

Dr. Raymond E. Janssen, a geologist, also took a great interest in marker trees and published articles on the subject in 1934, 1938, 1940, and 1941. In the 1940 *Natural History* article, "Trail Signs of the Indians," he notes:

> Yes, even now we may still see old Indian trail signs (trees) in many places. They are most numerous in the region about Chicago, but may also be seen in various localities throughout the Mississippi Valley, in Texas, and in the Great Smoky Mountains and Pocono Mountains in the east. I have seen them in Southern Illinois, Michigan, Wisconsin, Ohio, Indiana, Kentucky, Tennessee, Missouri and Arkansas. . . .
>
> In bending, the main stem of a young sapling was pulled downward and tied in position with a tough vine or strip of rawhide. The tree was always bent so that it pointed parallel to the trail. After a time the fastening would rot away. By then, however, the tree was permanently deformed. . . .
>
> These trail signs are the result of the Indians' custom of bending saplings and fastening them in position so that the direction of the bend indicated the route to be followed. . . .
>
> Occasionally when no young tree happened to be growing in a spot where a trail marker was desired, the Indians resorted to the bending of the lowermost branch of an older tree. The effect upon that particular branch was similar; the branch put forth new side branches which extended upward at an odd angle from the main one. (116, 117)

It is interesting that Janssen notes finding Indian marker trees in Texas, among many other states. He seemed to understand a great deal about the subject for his time, and a part of his knowledge may have come from earlier authors. Those interested in the subject gathered information and carried forward what was known in the past. Today, too, the previous research by others helps us to better understand how Comanche Marker Trees may be similar to or differ from those in other parts of the country.

When no sapling was available. Illustration of photo by Raymond E. Janssen

The Process of Qualifying and Recognizing Trees

CHAPTER 3

STEVE HOUSER
WITH JIMMY W. ARTERBERRY
AND LINDA PELON

The process used to qualify and officially recognize Comanche Marker Trees was developed over many years and is based on respect for the rights of the Comanche Nation, the property owners, as well as the trees involved. When our research began, there was no established protocol or set of procedures to follow. This book is the product of much research and, importantly, careful listening to our Comanche friends. Our process helps to protect the rights of property owners, does not injure any tree, and is based on a profound respect for the American Indian culture as well as their way of life.

In 1997, the Comanche Nation officially recognized its first tree—in Gateway Park, located in an East Dallas neighborhood. Since that time, a great many conversations and meetings have occurred to help us understand more about Comanche culture and how it relates to these important trees. Over time, researchers developed a process for qualifying trees that were submitted as well as respectfully requesting the Comanche Nation's official recognition of marker trees that appeared to be qualified.

Submission of Potential Indian Marker Trees

The first step of the process involves the submission of a potential Indian marker tree to the coalition for consideration. Anyone may submit a tree for consideration. We happily receive information on potential Indian marker trees from knowledgeable Comanche sources, foresters, ethnohistorians, anthropologists, geographers, master naturalists, community leaders, and hikers, as well as tips from people who simply share a love of these trees. A submission is typically made via the coalition Web site and usually includes a photo and a few details.

At this point, the information is reviewed to determine if the tree has potential to qualify. Sometimes, natural forces—rather than American Indians—clearly shaped the appearance of the tree. Trees that continue to have potential will require more research, so a file is created for "trees submitted." The person submitting the tree is asked to provide more information about the tree, such as its size, the property owner's contact information, and much more. Once the additional information is received, it is reviewed to determine if a tree (or trees) still has potential to qualify. At this point, the tree or group of trees is rated according to whether (1) it still has potential, (2) something rules it out, or (3) its status is not clear without an onsite investigation and further research.

Official File Created

A site containing trees that qualify is provided a number, and if there is only one tree on the site, the number represents the site and tree. If there is more than one tree, each tree is assigned a letter. For example, site 125 could have trees 125a, 125b, and so on, all listed in the file. There is also a letter next to the site that helps in tracking its status: "R" means the tree is officially recognized, "C" means the inspection is complete, "TD" means the inspection is still left to do, and "X" means the tree has been ruled out or ruled out until further facts or evidence is provided.

Establishment of the Indian Marker Tree Committee

After the inspection of a considerable number of trees, it became apparent to the coalition that there were far too many trees for a handful of people to inspect and research. As a result, in 2011 a team of volunteers was established, comprised of master naturalists, master gardeners, citizen foresters, community leaders, historians, and others with pertinent backgrounds. Each person is taught what to record, collect, or gather, and what basic procedures to use. As with many nonprofit groups, volunteers are asked to sign a release of liability due to the fact that they work on property owned by others. In order to protect the rights of property owners and the coalition, volunteers also sign a confidentiality agreement. This offers some level of guarantee to property owners that information gathered will be kept confidential and used only for the purpose of seeking official recognition of the trees.

The team has developed into an enthusiastic group of volunteers who travel at their own expense to help our efforts. As volunteers, all of us tend to measure our wealth by the good will built with others around us rather than in dollars. We know that a tree we inspect may not turn out to be old enough, or it may be ruled out in the future. We recognize that each site and each person we meet is different and that we can never know whether the trees and people we visit prove worthy of a chapter or a note in a future book.

Qualifying Factors

Submissions to us must be analyzed to determine which trees have the potential to be the genuine article. We developed a comprehensive list of qualifying factors that serve as guidelines to help us distinguish an Indian marker tree from other trees.

According to Comanche Nation Tribal Historic Preservation Officer Jimmy W. Arterberry:

> The Comanche entered into Texas during the early part of 1700 and by 1750 had gained full control of the lands as far south as the Edwards plateau. Due to the historical timeline, a tree in the northern parts of Texas could not have been formed before 1700, and a tree in the southern parts could not have been formed before 1730. As a result, trees in the northern region of Texas cannot be over 315 years old (2015 minus 1700). Trees in the southern region of Texas cannot be over 285 years old (2015 minus 1730).
>
> The signing of the Medicine Lodge Treaty occurred in 1865, and the last reference to free roaming Comanche occurred in 1871. This treaty established the Fort Sill Reservation, which is where the Comanche Nation is based today. Some individuals within American Indian tribes were known to have stayed behind when tribes were forced to move. However, there is no evidence to date which suggests that those staying behind either maintained existing trees or bent new trees. It would be most probable that the landscape had already been fully flagged by these types of trees by 1871. As a result, the minimum age for a Comanche Marker Tree is 144 years (2015 minus 1871). All potential Comanche Marker Trees must be between 144 and 315 years old in North Texas and between 144 and 285 in South Texas.

A Comanche Marker Tree is typically a native tree species that is known to be long-lived for the given part of the state. Without a doubt, the Indian tribes were quite knowledgeable about the life expectancy and cultural habits of native tree species, as well as other plants. As a result, the tribes often carried seeds of important plants and trees as they moved to new locations. What we see today as native species of plants and trees may not have existed without American Indians and the wildlife that moved them around many years ago. Animals and birds, of course, can eat the seeds of some plants and drop them in new locations. Or if trees and plants are on a hill or a mountain, the seeds might spread downhill. Some trees and plants produce seeds that are spread by the wind or other natural forces. Over time, all these actions can expand the native range of a species. Bearing all that in mind, when we find an old tree that is a nonnative species to an area, we always question where it came from and how it got there. A non-native or really out-of-place species may add to the case that a tree could be a genuine marker tree.

Comanche Marker Trees often feature sharp bends in the trunk. To bend these trees would have required a keen knowledge of the biological function of a tree's vascular system. To create the bend, often called a "hip," on some tree shapes may have required the removal of bark and underlying tissues. As a result, it is common to find old wounds and scars on the trunk, especially near the base or sharp bend. Scars and wounds in the right places can help to qualify a tree, but the lack thereof does not disqualify a tree.

Comanche Marker Trees are often associated with natural features significant or important to the Comanche that still exist today or may have existed many years ago. Trees were often markers for significant natural features, such as a good place to cross a river (low-water crossing), features on a high point (smoke signaling station, lookout point, paint rock quarry, and perhaps a place to leave messages for other members of bands, war parties, etc.), or a natural spring. These are a few obvious examples, but to find the purpose of a tree often requires "reading the natural landscape." The phrase describes the need to look at everything—the topography nearby, significant natural features, even existing and previous tree species and plant communities (or ecosystems) in a given area. When we first start to investigate a tree, the purpose of the tree is seldom ever clear. Studying all the natural features that exist around a tree today might not provide an accurate picture of what existed many years ago. Further investigation often leads to an answer, but to truly read the landscape, we need to rely on the knowledge of those who are familiar with Comanche culture and heritage. Essentially, we must look very closely at the landscape through the eyes of a Comanche Indian. A Comanche Marker Tree always had a purpose, but not necessarily one related to a significant natural resource.

Comanche Marker Trees are often associated with previous eyewitness reports of an American Indian presence in the area, as well as records indicating that arrowheads or other artifacts have been found on the site. Keep in mind that eyewitness reports should be substantiated or well documented rather than mere hearsay. Also bear in mind that it is not uncommon for someone to find an arrowhead on a site, but this alone does not establish a direct link between the Comanche tribes and the site. Frequently, sites known to be Traditional Cultural Properties have been researched by archaeologists and others with relevant educational backgrounds. Their findings offer an important qualifying factor for a tree(s).

We expect Comanche Marker Trees to be at least 22 inches in trunk diameter at or near the base. Although it may be possible to find a smaller tree that could qualify, this is the measure of the smallest tree to date that is believed to be old enough to qualify in Texas. The overall height or crown spread of the tree is not of great importance as a qualifier, although it does tell us a great deal about the past growth of the tree and its surroundings.

Comanche Marker Trees often have shapes or forms that have been formally recognized by the Comanche Nation. In addition, there are some shapes currently under investigation that may expand our knowledge of shapes that were commonly used in the past.

Site Visit and Inspection

If a site appears to have one or more trees with potential, we work with the property owner or sometimes with the person submitting the tree to schedule a site visit and tree inspection. In some cases, the tree may be located on public property, which does not require permission. Our volunteers always spend time with property owners or those submitting trees, in search of facts about the site and its history, as well as any pertinent facts regarding the specific background of a tree.

The inspection includes photos taken from all angles, as well as close-up shots of any scars, wounds, unusual bark patterns, or appurtenances found on the trunk. These may or may not be of significance, but the intent is to record everything we find that could be used as supporting evidence for the tree's qualification or for its age projection. Old cut wounds or dead limbs are analyzed, as noted in chapter 5, "Determining the Age Range of a Tree." Various size measurements are recorded, as well as GPS coordinates and the direction of the point or lean, assuming one exists.

After a file on a tree is created and after the site visit is complete, a tree is either deemed, at that time, eligible to qualify or it is ruled out. Even if it does not currently qualify because of age or other factors, any future submission of facts or evidence could potentially change the status of a tree. A tree previously ruled out can be reconsidered if significant information is later uncovered.

Background Research

After the site visit and inspection, if the tree is old enough and still has potential to be an Indian marker tree, a great deal of background research is required. The objective is to make a case that Comanche were likely to have occupied the site containing the tree(s) around the time it was originally bent. This can be a difficult task, and it often requires determined sleuthing. There is a large and relatively unexplored paper trail of documents, reports, captive narratives, and additional primary sources available to help us establish American Indian or Comanche presence in many of the areas where we find potential marker trees. As a result, research may be supplemented with a search of property records at the county courthouse and progresses to checking with local historians, anthropologists, and archaeologists, among other knowledgeable individuals.

Although each tree with potential will be inspected over time, there are simply not enough volunteers with an interest in fully researching the individual sites or trees. The coalition chairperson for the Historic Research Committee, James Alderman, is an excellent research analyst, but there are many more trees than researchers. To date, more than 450 trees have been submitted for investigation, and data have been collected and filed on over 150 trees. The Historic Research Committee must also provide resources and data for the Historic Tree Committee, chaired by Bill Seaman.

The process relies heavily on our volunteers and people in their communities around the state to help us research the potential past presence of

American Indians near a potential marker tree. The information provided is reviewed, and any resources listed are verified. Our individual researchers often find significant new information because we are looking for a different type of data than that sought by other researchers.

Beyond researching the age and details of a tree, we also research the distant as well as the recent history of a potential Indian marker tree site and the surrounding area to find any documentation that may help to quantify American Indian occupation there. We also look for documentation of a site and its trees in general, studying any odd-shaped or large tree noted in an old property survey, or examining old photos of a property that may show trees in the background. Old family photos or photos stored in various historical archives may contain images or drawings of trees. Artwork or postcards from the past may contain trees, as well.

This type of research can be done by a novice, but having a historian, anthropologist, archaeologist, or other professional involved can help immensely. They often know where to look and what to look for. It is good to record oral information based on what people in an area may remember, but good published documentation holds more weight in verifying the history of a tree or an area.

It is worth noting that none of the information we currently seek was available online in the past, while new information becomes available every day. Likewise, greater amounts of information may be available in the future, and some of the information contained in this book can now be found online.

Because some trees in which we are interested pointed direction to travelers, we often seek information that can be found on early trail maps of an area. We also look at current topography maps to determine how a tree may have related to the earlier landscape. An area of low elevation near a river, creek, or spring may contain a tree marking a campsite, whereas a high point may mark a lookout point or a smoke signaling area.

In our recent conversation with a person submitting a tree for consideration, a question arose regarding the direction of the point, and the property owner was asked to look at the topography in the vicinity. She explained that the tree points to a large hill in the area but that there is nothing on the hill. We asked her to go back and look on the ground for rocks that might not appear to be from the area. Deeply burned rocks may be evidence of hearths that were used for smoke signaling. Stone tools or chert flakes often provide evidence of the tool-making process. Ochre, black, yellow, and other colored rocks used as paint pigments may also indicate a Comanche presence.

She called back to explain with great enthusiasm that she had found many uncommon rocks not native to the area. It was gratifying to realize that we are uncovering parts of history with our work that might never be recognized otherwise.

Current satellite maps from a number of resources also aid our research. If you use these maps to look at a tree out in the open or in a lower-density forest during the dormant season, you can often see the bent trunk of a tree. As we study early trail maps, we can use these satellite maps to look for potential Indian marker trees on private property. Advancement in technology will undoubtedly provide more resources in the future. An inventory of trees conducted in 2008 using cutting-edge technology included 20 square miles of the forest canopy in Dallas, generating high-resolution pictures of the forest that provide another resource for finding these trees on the ground, this time from an airplane rather than a satellite.

Many maps can be found at the Texas General Land Office, local libraries, and universities, as well as through Internet sources. Useful documents can include early hand-drawn maps of missions, forts, colonies, water resources, and battle sites, among others. *Texas Indian Trails* by Gelo and Pate is another primary resource on the subject of Comanche trails, listing some of the forts and other natural resources along the way.

Often many types of animals created the earliest of trails as they migrated, seeking food, water, or shelter. As you would expect, these trails were also followed by American Indians and ultimately, early settlers. It is surprising to find the number of current roads that follow a part of an early trail because the way was already cleared or the path made sense from the standpoint of an early traveler negotiating a difficult landscape. Trails often proceeded from animal use to human use and then to cars.

As we examine rivers, tributaries, and creeks, we always look for changes that may have occurred over time. A tree that was once near a water feature may now be quite a distance away because of alterations in the watershed. By the same token, we suspect that a few trees that once marked the way to water may now be at the bottom of a lake built as part of recent watershed development.

Media news reports from the past century also often contain important information about the history of a site, maps of natural resources, details of infrastructure as it developed, and more. Unfortunately, archives of these early materials are often inaccessible or difficult to search.

Early property surveys often referred to trees as marking points on a property. Old property deeds often show the genealogy of those who once resided on a property. These can be found in local county courthouses or possibly the Texas General Land Office's History and Archives section (http://www.glo.texas.gov/what-we-do/history-and-archives/index.html). When previous occupants of a property can be located, it is possible that old family photos will show a tree in the background, providing more history about a tree and its size at some point in the past.

As part of the big picture, the Texas Historical Commission serves as a clearinghouse for many of the facts and documents we seek and offers support in various areas of research. Other resources include the National Register of Historic Places, the Office of the State Archaeologist, and others, as well. (Link: http://www.thc.state.tx.us/).

Here are a few other helpful resources:

• *Fundamentals of Oral History: Texas Preservation Guidelines* is a handbook for collecting oral history which can be printed from the link below. It is an excellent guide for learning how to collect oral histories. The publication includes suggested forms for obtaining the permission of the interviewee to share the information.
    (Link: http://www.thc.state.tx.us/public/upload/publications/Oral History.pdf)
• There is a great deal of information on the Texas Historical Commission's Web site that is helpful in collecting data about trees under investigation. The information on State Archaeological Landmark Designation is useful for recognizing archaeological sites, including archaeological footprints of previous land uses.
    (Links: http://www.thc.state.tx.us/about/our-divisions and http://www.thc.state.tx.us/public/upload/archeology-state-landmark-desig nation.pdf)
• *Guidelines for Evaluating and Documenting Traditional Cultural Properties* is a bulletin available online from the National Register of Historic Places. This publication from the US Department of the Interior provides more helpful information on Traditional Cultural Properties.
    (Link: http://www.nps.gov/nr/publications/bulletins/pdfs/nrb38.pdf)

Local or regional libraries often house historical documentation regarding the early history of an area. Check for historic markers in your county or area, and follow up with visits to local libraries. They often contain valuable information regarding the presence of Indian tribes.

University libraries are repositories of useful information. Historians, anthropologists, and archaeologists often have easy access to many important documents that might not be available to the public.

We provide these resources for those who may wish to research the subject, a site, or an individual tree. Old maps, property surveys, or previous recorded history of a site may provide help in qualifying a tree as a Comanche Marker Tree. It is important to establish a link between a site and the past presence of the Comanche.

Official Recognition

As a part of our process, once there is substantial evidence and research linking the tree and the site to past American Indian presence, the information is submitted to officials of an American Indian nation. In one instance involving historic trees, we worked with the Waco/Hueco, Wichita, and affiliated tribes to formally recognize a grove of historic live oaks in Waco, Texas. The grove grows along Barron Branch Creek, a tributary of the Brazos River near downtown Waco. The grove is old enough to have been a part of the Waco/Hueco culture, and the trees were undoubtedly witnesses to the tribe leaving the area for reasons that remain unknown. Once the site and live oak grove were linked to the past presence of the tribe, the Texas Historic Tree Coalition issued a formal proclamation about the grove, along with a handsome, framed certificate.

As far as Indian marker trees are concerned, we work primarily with the Comanche Nation, the only group that recognizes the existence of these trees in Texas. Once the information about a tree is submitted, many questions can arise, requiring a great deal of time and effort to investigate further. Any official response regarding a tree is respectfully left in the hands of Comanche officials. It is important to note that it is a slow process due to the many other matters these officials must address.

Our coalition has a great respect for the history that is passed along through oral tradition in the American Indian culture. Although there is very little written information available to the public, we find that the oral

A framed certificate provides formal recognition of a tree or grove that is suitable for any office.

A formal proclamation explains the history and importance of a tree or grove.

traditions of American Indian tribes have preserved their history, culture, and values exceptionally well. Their tradition of preserving and passing along aspects of their culture has produced generations of people that know who they are, what they believe, where they have been, and where they are going. Contemporary Comanche know that some of the trees we submit are the same as those they heard mentioned by elders as they grew up. We highly value and respect the decisions they make about the trees that we submit for consideration. Comanche created these trees, and our process rightfully ends with them. Once the Comanche officially recognize a tree, the decision is beyond reproach unless new and significant facts are uncovered.

Volunteer researchers, professionals, and experts in various fields continue to study and research these trees to determine which ones have potential. An odd or bent shape has meaning only when research shows the tree to be old enough and when it possesses many of the traditional characteristics found to be associated with Indian marker trees. Although this research is important to verify the potential of a tree, only an official American Indian tribal representative can declare a tree to be a part of that tribe's cultural heritage. The difference between any bent tree and an Indian marker tree is the research and science behind it, plus recognition by the proper authorities. Once recognized, these trees bring their unique forms to the larger family of historic trees.

We and others working around the country all recognize that an American Indian tribe that once existed in an area may no longer exist. In these cases, combined research can help us reach a higher degree of accuracy in determining which trees are legitimate Indian marker trees. Without the presence of an American Indian tribe to recognize a tree, it is important to gain insight from others with experience or knowledge on the subject.

Project Challenges

In the beginning of our effort to recognize Indian marker trees, one of the primary challenges involved was the significant lack of information. With very limited information or data ever produced on the subject, it was difficult to verify the trees' existence. The early hope was to find previous research or recorded oral history relating to these trees in Texas. There are many more primary sources to explore and oral histories to collect that should shed new light on this important topic. Although American Indian officials verified the existence of some trees in a few parts of the country, none had ever been officially recognized in Texas.

It became clear that a great deal of research was required. Much of ours was done behind the scenes over the years. We wanted to have enough data to be comfortable bringing the topic to the public. What turned the tide for our research was the continued support from Jimmy W. Arterberry in verifying our findings, providing information, and working toward recognition of these trees.

Another significant challenge continues to be working with various property owners to explain our mission, obtain permission to record data, and yet still maintain their privacy. Property owners are typically concerned about the information gathered and its potential future use. An explanation of the process we use is required, including the fact that all information is kept confidential. A requirement to join the coalition's volunteer efforts involves signing a release of liability but also a nondisclosure agreement that helps to protect a property owner's rights. Property owners can sometimes have concerns that recognizing a tree may limit their use of the property or create some restrictions. This is simply not the case. Recognition of the tree(s) is not the same as a historic site designation, Traditional Cultural Property, or other designations that may affect the use of the site. An owner is not required to care for or protect the tree, but we do usually request notification if the tree is lost for any reason.

Managing the Indian Marker Tree Committee's visits to sites can be quite a challenge. Some sites can be accessed easily, but others can require an off-road vehicle and a lengthy drive or walk. This requires packing all the required tools, preparing for a longer trip and sending volunteers who are not afraid of snakes or spiders. We also must consider the weather. Winters in Texas may not seem bad, but one trip required a 30-minute ride on an open four-wheeler when the wind chill was in the single digits. On one previous summer trip to Austin, the temperature was nearly 100 degrees in the afternoon. We also must coordinate volunteer efforts with a property owner's schedule. Thankfully, team members Monica Morrison and RuthAnn Jackson help in this regard.

As with any volunteer nonprofit group, those involved do not always agree or may have different agendas than that of the primary group, which can add to the challenges of a project. It is not uncommon to see a non-profit struggle or cease to exist due to internal conflicts. Recognizing those who have conflicting agendas and resolving the issues can be difficult at times. Most want to enjoy their work as volunteers, and they want to see results as well as the benefits of their efforts. When there is not a cooperative effort to follow a unified vision of the future, volunteers can start to lose interest. The coalition is well aware of these types of challenges and works diligently to resolve problems before they become a greater concern.

The fact that we deal mostly with confidential information that should never be exposed to the public or the media adds to the challenge. It is not uncommon for a property owner to request media attention when a tree is being researched and before it has been officially recognized. Although this is always discouraged, media attention does happen. Our work is off the record until a tree is officially recognized. It stays off the record even after recognition, if requested by a property owner. We believe that trees currently being researched should not be given exposure until our work reaches a conclusion.

Comanche Nation Management Structure and the Role of the Tribal Historic Preservation Officer

BY JIMMY W. ARTERBERRY

The Comanche Nation is governed by a constitution and by-laws, which delegate the authority of an Elected Business Committee with the responsibility to appoint or hire individuals to perform certain duties, and manage all other tribal affairs as outlined in the constitution. The Comanche Nation Business Committee (CBC), which is comprised of a chairman, vice chair, secretary/treasurer, and four members-at-large, creates laws in accordance with the constitution. One such ordinance establishes a Tribal Historic Preservation Officer (THPO) and outlines the duties and responsibilities of that individual.

In addition to performing routine technical review and compliance activities under the National Historic Preservation Act (NHPA) of 1966, as amended, the THPO works closely with various other governmental officials, scholars, researchers, artists, educators, students, elders, and tribal members on Comanche history matters. Generally speaking, the THPO must be well versed in all Comanche history matters and preferably a member of the Comanche Nation as well. The THPO is an appointed and hired official who has been charged by tribal law and endorsed by the National Parks Service (NPS), via a contractual agreement, to perform the duties described. In essence, the THPO is the official entity within the Comanche Nation's governmental structure that reviews cultural data internally and externally for protection, endorsement, and/or denial purposes. Furthermore, the THPO is the key support of the CBC on all legal matters where Comanche culture and history are in question or challenged.

Where the investigation of a potential Comanche Marker Tree is concerned, the THPO works very closely with investigators to provide guidance and make determinations on the authenticity of a tree. However, it must be noted that the tribal chairman, too, can make such assessment, without review or discussion with the THPO. That is known as "a reserved right" granted to the chairman by federal rule.

Applying the Landscape Model: Understanding Comanche Marker Trees

BY LINDA PELON

One of the most challenging aspects of recognizing Comanche Marker Trees and sacred sites is that material evidence of Comanche occupation often is not present at these sites. Artifacts documenting historic Indian presence can be completely removed by surface collecting and natural forces such as flooding and erosion. This historic level is the very thin upper level of a site, and more than 150 years of exposure can dramatically change or destroy this archaeological evidence. Also, Comanche environmental ethics demand a very light human touch at sites considered sacred. These sites are used for spiritual purposes that did not have a significant impact on the land, and respect for these places resulted in a careful cleansing of them after use. Therefore, it is often necessary to find other methods of verifying Comanche land use. Traditional knowledge preserved in Comanche oral histories, stories, and other aspects of cultural heritage can be combined with "reading the landscape" as an effective strategy for recognizing Comanche land use patterns. This *landscape model* is recommended by the NPS for nominating sites to the National Register of Historic Places as Traditional Cultural Properties. The NPS bulletin *Guidelines for Evaluating and Documenting Traditional Cultural Properties* explains this process:

> One kind of cultural significance a property may possess, and that may make it eligible for inclusion in the Register, is traditional cultural significance. "Traditional" in this context refers to those beliefs, customs, and practices of a living community of people that have been passed down through the generations, usually orally or through practice. The traditional cultural significance of a historic property, then, is significance derived from the role the property plays in a community's historically rooted beliefs, customs, and practices.
>
> Traditional Cultural Properties are often hard to recognize. A traditional ceremonial location may look like merely a mountaintop, a lake, or a stretch of river; a culturally important neighborhood may look like any other aggregation of houses; and an area where culturally important economic or artistic activities have been carried out may look like any other building, field of grass, or piece of forest in the area. As a result, such places may not necessarily come to light through the conduct of archaeological, historical, or architectural surveys. The existence and significance of such locations often can be ascertained only through interviews with knowledgeable users of the area, or through other forms of ethnographic research. The subtlety with which the significance of such locations may be expressed makes it easy to ignore them; on the other hand it makes it difficult to distinguish between properties having real significance and those whose putative significance is spurious. As a result, clear guidelines for evaluation of such properties are needed.

Instruction on using the landscape model was provided by Dr. Richard Francaviglia, director of the University of Texas at Arlington's Center for Southwestern Studies and the History of Cartography, when he became interested in my work on Comanche land use and invited me to apply for the graduate research assistant position for the center he directed. Dr. Francaviglia, a prolific author skilled in using this new model called "reading the landscape," became my mentor and trained me to apply it to Texas Comanche land use. The use of this model became the foundation for continued research on Comanche Marker Trees and the identification of sacred sites. The skill set involved in this research included understanding land use from a Comanche perspective. I have been very fortunate to have the assistance of Comanche elders, artists, historians, and tribal leaders who generously shared information and trusted me to use it in a manner that is helpful to their nation. Supporting information must also be acquired through the historic record in order to document and map Comanche trails, campsites, lookouts, signaling points, rock art sites, significant environmental resources, and other information used to place suspected traditional cultural properties into a larger environmental and historic context. Site visits, surveys of archaeological and historic records, collection of oral histories, and other methods are also used to recover information and evaluate the significance of suspected Indian marker trees and other Traditional Cultural Properties.

First, a nominated Indian marker tree must be examined by a professional arborist in order to verify that the tree cannot be ruled out by age or other criteria used by arborists and foresters. The next step is to conduct ethnohistorical research to recover information to verify Comanche occupation and use of the area that includes the tree. At this point, the tree may become a sort of storyteller, since important and neglected information about Comanche land use in that area is brought into focus and clusters around the tree. The recovery of this information can reveal aspects of Texas Comanche history never before recovered, organized, and documented. This often leads to insights that contribute to a more nuanced understanding of the historic Comanche presence in Texas.

The following summaries of a few of these Comanche Marker Tree investigations are provided to help readers appreciate the importance and validity of using the landscape model to document Comanche Marker Trees and other Traditional Cultural Properties. With the recent creation of an ethnographic taxonomy of Comanche categories for trees, the definition of Comanche Marker Trees was expanded to include some trees that are not bent. Earlier research was focused only on bent trees. There are many trees in various stages of investigation that are not discussed here. These bent trees have been found at low-water crossing, at intersections of major

Indian trails, at former Indian campgrounds, and near significant natural resources such as springs and a paint rock quarry. Research on Indian marker trees outside Texas also supports the hypothesis that these bent trees are landscape navigation aids or intentional signage, rather than trees that coincidentally occur at places significant to Indian land use.

Comanche Marker Tree Taxonomy

Comanche Marker/Turning/Pointing/Leaning/ Bent Trees (Medicine Trees)

JIMMY W. ARTERBERRY

C H A P T E R 4

All trees can be marker trees, but are not. We must be clear that marker trees are meant to identify a location, regardless of event or resource. They are locating/calling/signaling trees that can give or deliver a message as well as direct a path. They can be guardians over something such as a hidden space. Marker trees can generally be described as landmarks, but are not always that, and also might be found within proximity to other landmarks. They are often tied to trails and other resources, but again are meant to identify location. They are mapping points and can sometimes appear as anomalies in a location, or they may be specific to a location (e.g., a storytelling tree).

Turning trees (which are also called pointing, leaning, and bent trees) are shaped or molded to indicate a direction. However, they can also indicate a resource or resources, or be a multiple resource themselves. They are marker trees and service trees. They can also be treaty/council trees, medicinal trees, ceremonial trees, or burial trees.

Burial trees are meant to be used as just that. They are also service trees, ceremonial trees, and marker trees. They can also be turning trees because of the location, as well as medicinal trees or treaty/council trees. Because of their nature, they are remembered for gathering times and serve as a healing source.

Ceremonial trees offer a service and are utilized ceremonially. They are also medicinal trees, burial trees, and service trees. Additionally, they can be marker trees, treaty/council trees, or turning trees. Generally speaking, they are used for ceremonial activities.

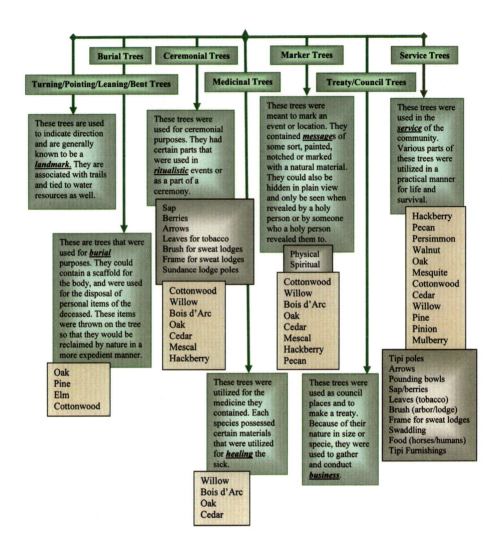

Burial Trees

Ceremonial Trees

Marker Trees

Service Trees

Turning/Pointing/Leaning/Bent Trees

Medicinal Trees

Treaty/Council Trees

These trees are used to indicate direction and are generally known to be a *landmark.* They are associated with trails and tied to water resources as well.

These trees were used for ceremonial purposes. They had certain parts that were used in *ritualistic* events or as a part of a ceremony.

Sap
Berries
Arrows
Leaves for tobacco
Brush for sweat lodges
Frame for sweat lodges
Sundance lodge poles

Cottonwood
Willow
Bois d'Arc
Oak
Cedar
Mescal
Hackberry

These trees were meant to mark an event or location. They contained *message*s of some sort, painted, notched or marked with a natural material. They could also be hidden in plain view and only be seen when revealed by a holy person or by someone who a holy person revealed them to.

Physical
Spiritual

Cottonwood
Willow
Bois d'Arc
Oak
Cedar
Mescal
Hackberry
Pecan

These trees were used in the *service* of the community. Various parts of these trees were utilized in a practical manner for life and survival.

Hackberry
Pecan
Persimmon
Walnut
Oak
Mesquite
Cottonwood
Cedar
Willow
Pine
Pinion
Mulberry

These are trees that were used for *burial* purposes. They could contain a scaffold for the body, and were used for the disposal of personal items of the deceased. These items were thrown on the tree so that they would be reclaimed by nature in a more expedient manner.

Oak
Pine
Elm
Cottonwood

These trees were utilized for the medicine they contained. Each species possessed certain materials that were utilized for *healing* the sick.

Willow
Bois d'Arc
Oak
Cedar

These trees were used as council places and to make a treaty. Because of their nature in size or specie, they were used to gather and conduct *business*.

Tipi poles
Arrows
Pounding bowls
Sap/berries
Leaves (tobacco)
Brush (arbor/lodge)
Frame for sweat lodges
Swaddling
Food (horses/humans)
Tipi Furnishings

Comanche Marker Tree taxonomy.

Medicinal trees offer materials for the healing of the body and mind. They are also service trees and ceremonial trees. They can also be marker trees, treaty/council trees, burial trees and turning trees.

Treaty/council trees are utilized for business purposes. They are service trees, marker trees, and ceremonial trees. They can also be turning trees, medicinal trees, or burial trees, but are generally denoted for their use in marking a particular location.

All trees are service trees and can be utilized in a variety of ways. They can also be used as marker trees, turning trees, burial trees, ceremonial trees, medicinal trees, and treaty/council trees.

Recognized trees must be between 144 and 315 years old to qualify as Comanche Turning/Pointing/Leaning/Bent Trees (which are a form of marker tree). These trees could also be burial, ceremonial, medicinal and/or treaty/council trees.

There are certain changes in tree uses since the reservation era, and these are some examples:

- Today, the Comanche do not utilize burial trees. However, there are trees being planted at gravesites today.
- There are treaty/council trees being used. One example of such a tree resides at the Comanche Nation Tribal Complex. The tree was used during the late 1990s as a gathering place to elect a new council (Tribal Business Committee) when tribal members were at odds with each other.
- Marker trees are still used today and are generally identified by location, such as the one referenced above.
- Ceremonial and medicinal trees are still used today, as well.

Tree Names in Comanche

The generic Comanche term for a trail tree is *nayuru hupi*. Each one can be or would have been additionally described by its species, but not always. It could have simply been called a trail tree, a marker to lead people to a particular place.

The Comanche term for any tree is *huupi*, and the type or species could be applied to the term. Depending upon who was describing the tree or giving directions to locate the tree, additional word descriptions would or could be used.

Rita Coosewoon has provided these Comanche names for the various species and has detailed the uses of each:

- Persimmon—*Naseeka* (fruits eaten fresh or dried and mixed with meat into patties)
- Pecan—*Nakutabac* (nuts eaten, leaves used to treat ringworm)
- Hackberry—*Nats kwi* (fruits mixed with fat)
- Dogwood—*Paruabi* (stems used to make arrow shafts)
- Thornapple—*Tidiam woo* (fruits eaten, also called "red haw")
- Thornapple—*Tupuk wu kwekate* (fruits eaten, inner bark chewed as gum)
- Walnut—*Mu itai* (nuts eaten, leaves used in treatment of ringworm)
- Juniper, Red Cedar—*Ekawapi* (fruits eaten, smoke from leaves used for spiritual purifying)
- Red Mulberry—*Etihup* (fruits eaten)
- Mesquite—*Nats kwe* (meal from the pods eaten, leaves used to neutralize stomach acidity)
- Wild Plum—*Kusi su kwi* (fruit eaten either fresh or dried and stored for later use)
- Blackjack Oak—*Duhup* (acorns eaten, leaves used as cigarette wrappers by elders)
- Oak—*Pesap ni* (acorns eaten, trunks used for fire wood and to build shelters)
- Willow—*Ohash* (ashes from fire used to treat sore eyes)
- Mescal—*Ekapu* (beans used as ornaments or as necklaces for ceremonial purposes)
- Osage Orange or Bois d'Arc—*Aha hupi* (bows made from branches, infusion of roots used in treating sore eyes)

(Note: There are a number of other names and uses for these trees as well as those provided by Ms. Coosewoon.)

Determining the Age Range of a Tree

STEVE HOUSER

CHAPTER 5

B y including information about how we determine the approximate age of a tree, we hope to provide a better understanding of the methodology we use, as well as the noninvasive and respectful way in which our research is conducted.

According to *Merriam-Webster's Dictionary*, *dendrochronology* "is the science of dating events and variations in environment in former periods by comparative study of growth rings in trees and aged wood" (1994, 308). As early as the fifteenth century, the relationship between rainfall and tree growth rings was documented. Over time, it became clear that tree growth rings also relate to a site's climatic zone, soil temperatures, biological activity in the soil, growth conditions in the root zone, and much more. Experts in the field have produced a great deal of research on the subject; the following text has been developed to provide only a basic understanding. For each year that a tree grows, it produces two new sets of straws, or growth rings. One set is developed in the spring and one in the late summer or early fall (sometimes referred to as *earlywood* and *latewood*). Each set of growth rings (or set of straws) is typically a different color, with one lighter in color than the other. Earlywood is lighter in color with latewood being darker. Counting either the light- or dark-colored growth rings on a stump or limb from the center of the rings to the outer edge will provide the tree's approximate age, the exception being false or missing growth rings. The number of rings found per inch in diameter will typically vary. Tree ring details can be studied and researched using a magnifying glass, microscope, machine, computer program, image analysis, or X-ray.

Conditions Affecting Tree Growth Rings

The distance between the growth rings is an indication of a tree's health, as well as the cultural or environmental conditions during each year, or years, of its life. Years of higher rainfall may produce larger straws or growth rings, whereas years of drought may produce smaller straws. The National Weather Service can corroborate a great deal of tree growth ring data, primarily because the size of the growth rings is directly related to past weather conditions and droughts for a geographic area.

The cultural conditions of a site relating to a tree such as past pruning, irrigation, or root/trunk damage, will also have a direct effect on a tree's health and growth rates. If a tree is growing on shallow rock, in poor soil conditions, or suffers from a lack of sun or other environmental conditions, its growth rates can be greatly reduced. Trees growing in a lower-lying area or near a creek will typically enjoy better soil and grow at faster rates than trees growing on shallow rock.

Growth rates can also vary greatly by species. Those species known to be faster growing, such as mulberry (*Morus* spp.) or cottonwood *(Populus* spp.)*, can produce large sets of straws or growth rings each year. Slower-growing species, such as post oak (*Quercus stellata*), can produce very small sets of straws, especially when growing on a rocky site with no irrigation.

Researchers at the University of Arizona's Laboratory of Tree-Ring Research and the University of Tennessee, Knoxville's Laboratory of Tree-Ring Science, among other professionals in the field, utilize a *tree ring growth model* to determine the age of limbs and trunks. The model is based on five primary factors:

- Age-related growth trends of nearby trees.
- Dendroclimatology or climate-related growth due to environmental factors, such as temperature, water, or atmospheric pressure.
- Dendroecology or disturbance-related growth due to local factors within a stand of trees, such as pollution, exposed roots, shaded location, or insect damage.
- Dendrochronology or disturbance-related growth due to regional factors, such as fires, floods, earthquakes, or other natural disasters.
- "Error" evaluations due to factors other than those noted above.

All these factors can have a direct correlation to tree growth ring size and structure. In essence, each tree has its own biological growth curve, which is expressed in the physical composition of the growth rings. As a result, tree size can be very deceiving when attempting to ascertain the potential age of an Indian marker tree.

Determining a Potential Age Range

As a general rule of thumb, determining the age of an upright and normally shaped tree requires taking a measurement at about 4.5 feet above the ground. This is commonly called *diameter at breast height* (DBH). The diameter of a tree at this point on the trunk can be multiplied by a very general growth factor developed for certain tree species to determine an approximate age range. A 40-inch DBH cottonwood would be multiplied by a factor of 2, to be about 80 years old. A 40-inch maple would be multiplied by a factor of 3.75, suggesting that the tree could be more than 159 years old. However, this is only a general method for determining potential age and does not take into account many variables or the fact that the tree could be bent.

While the only way to know the *exact* age of a tree is to cut it down and count the growth rings at the base of the tree, another estimating method used by professionals is to remove core samples of wood using an increment or displacement borer. The process involves drilling into the trunk of the tree to remove a core sample of wood and counting the growth rings. Unfortunately, the process can have negative effects. According to *Tree Rings: Basics and Applications of Dendrochronology* by Fritz Hans Schweingruber, "Damage can result mechanically from the borer itself, physiologically through the reaction of the tree or pathologically through foreign organisms" (1998, 44). Regarding the intensity of the damage, Schweingruber writes, "Now, while it is certainly true that resin-rich species, e.g., pines and some species with a high tannin content, are only rarely attacked by fungi after coring, there are nonetheless species in which coring does cause a certain amount of damage" (44). A tree's response to wounding can extend beyond the area of the core sample removal. In *The Vascular Cambium: Development and Structure*, Philip R. Larson notes, "The response to wounding is not confined to the immediate vicinity of the wound. It can extend quite some distance both radially and longitudinally from the wound site" (1994, 501).

Using this method on potential Indian marker trees introduces the following concerns.

- Reading or counting growth rings can often be difficult from a core sample of wood because the sample can fall apart during the removal process.
- Some tree species' wood fibers are not easily collected through coring, which can complicate the process and obscure the results (e.g., live oak, *Quercus virginiana* or *Quercus fusiformis*).
- The wound produced from coring the trunk can cause physical damage to the tree's vascular system, reducing its ability to function normally.
- Coring increases the potential for decay to spread in a very critical part of a tree—in the trunk or near the base of the trunk. It also leaves the tree open to infection by various pathogens.
- A typical core boring that penetrates the outer 16 inches of a large tree trunk indicates only its most recent growth rate, as opposed to the growth in its early years, which is recorded by the rings near the center of the trunk. Longer core boring extensions can be used to increase accuracy, but it also extends any potential negative effects farther into a tree.

Considering that the trees we study were potentially created by American Indians, we simply do not feel it is respectful to drill into them.

Another Method

Without core boring, we needed another way to document the potential age range of a tree without creating a wound. The method we use is grounded in basic measurements, but more on experience in dendrochronology and a thorough understanding of local tree species and their individual growth rates. As a result, using this method requires individuals be knowledgeable in these fields.

Rather than bore into a tree, we prefer to study the growth rings that might be found on an old cut wound or on dead limbs. Comanche officials have agreed that removing a dead limb is not disrespectful because dead-wood does not function to support the health of the tree and it will not regenerate.

Larson notes, "Branches of trees have often been the source of material for investigations of anticlinal divisions of cambial cells because sampling was both easy and less injurious to the tree. The patterns and consequences of anticlinal divisions in branch material differ little from those in stem material when consideration is given to rates of radial and circumferential branch expansion, age, eccentricity, and other contributing factors (Whalley 1950; Evert 1960, 1961, 1963a; Cheadle and Esau 1964). . . . As a general rule, xylem elements in both branches and roots are smaller than those of stems (Dinwoodie 1961)" (1994, 251). Larson goes on to say, "In view of the fact that branch growth is often eccentric, one might expect to assume that the differential growth on the upper and lower sides would influence both the frequency of anticlinal divisions and the sizes of the derivatives. Therefore, caution should be exercised when assessments of cambial divi-sion patterns for a species are based on branch material" (252). Although Larson urges caution, there is no other viable option for our research short of core boring.

In inspecting the tree, we locate the largest exposed cut wound in the lower sections of the tree or the largest dead limb. If a dead limb is used to "read" or view the growth rings, it is safely removed and cut into a thin section, often called a *tree cookie*. Depending on the species of tree and the characteristics or texture of the wood, the tree cookie or the wound on the tree may need to be sanded or stained to read the rings accurately. In some cases, applying water, tea, wood stains, or coffee to the cookie or wound will help to bring out the color of the growth rings.

The longer a limb has been dead, the harder it is to read the growth rings. This is due in part to the loss of moisture in the wood. It can be dif-ficult to read growth rings on some tree species or on limbs with decay. The more extensive the decay, the more difficult it is to read the growth rings.

In some cases, a magnifying glass is required to see and read the growth rings. Once the growth rings are visible, a ruler is placed across the wound or tree cookie and the number of rings per inch are counted in several places and recorded.

To count rings on an old cut wound found on a tree, we take a photo-graph of the wound with a tape measure stretched across it. This provides

Tree cookie before adding coffee as a stain. The site number is often marked on the tree cookie for identification.

The same tree cookie after the addition of stain to help in reading the growth rings.

Growth rings are smaller on one side than the other.

the number of growth rings per inch in various parts of the old wound. In some cases, rings are much tighter in some parts of the wound than in others.

The larger the wound or tree cookie, the more accurate an estimate may be of the long-term tree growth. The approximate age of the tree can be based on more than just a small number of growth rings when a larger wound or dead limb that shows a large number of growth rings can be found. If all the rings on a limb or trunk are visible, the exact age is clear, as long as any potential false or missing growth rings are also considered.

The lower an old cut wound on the main trunk may be, the more accurate we can be in determining the average growth rate of the tree. Wounds on limbs and tree cookies measured in the upper portions and outer canopy sections are more indicative of the tree's recent growth rate, whereas those removed and measured in the lower or interior sections of the trunk will normally show earlier growth rates.

It is important to note that most growth rings we are able to count represent only the more recent growth rate as opposed to the rate when the tree was younger and smaller. However, knowing the more recent growth rate provides important insight into a tree's potential previous growth rates.

For the purpose of determining a tree's potential age, we measure a tree just above the root collar, or root flare if one exists, and use a diameter tape (D tape) stretched around the trunk. (A regular tape measure, rather than the D tape, can be wrapped around the trunk to measure the tree's circumference. A cloth tape measure may be more accurate than the typical metal tape measure, which may not conform to the trunk, allowing kinks to form, which can cause measurements to be larger than the actual size.) To obtain the diameter, divide the circumference by 3.1415.

Determining Average Growth Rate

Since trees produce new straws on all sides of the trunk or branches each year, we count from the center of the growth rings to the outer edge of the limb, trunk, or stump to accurately determine the age. Although the center of the growth rings might not be in the exact center of the tree, we still use the radius to calculate the general age range. (To get the radius, divide the diameter by 2.)

To determine an average growth rate, measure the rings closer together for one average and rings with the widest spread for the other average. Add the two together and divide by two to get an average growth rate. If the center of the growth rings is not located in the center of the wound or tree cookie being measured, look for an average spacing rather than measur-

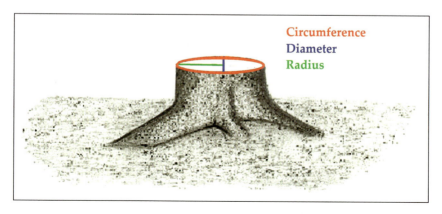

Basic measurements of a trunk or limb.

ing only the tightest rings or those that are farther apart. Measuring only tight rings or wide rings can lead to inaccuracy if the center of the growth rings is not in the center of the stump. An average growth rate simply adds another factor that helps in determining the potential age of a tree. Taking the average growth rate per inch in diameter and multiplying it times the radius will not accurately reflect the age because the growth rate often reflects only *recent* growth, which might or might not be the same as the growth rate in the earlier years of the tree's life. This is why the many other growth rate factors previously mentioned must still be considered.

Once an average or exact growth rate for a period of time is determined, it is compared to the size of the trunk at the base. Since the center of the growth rings may be asymmetric or not always in the exact center of the tree, the radius measurement is only an approximation. Many Indian marker trees have trunks that are oblong, rather than round. According to Schweingruber's *Trees and Wood in Dendrochronology*, "In coniferous species, site changes produce compression wood on the lower side. In deciduous species, site changes produce tension wood on the upper side" (1993, 5). These statements are in reference to site changes related to wind, snow, and unstable soil, which leave trees with a bow or angle in the trunk. The

Very small growth rings found on a bent Eastern red cedar (**Juniperus virginiana**).

reaction wood moves the center of the growth rings off to one side and creates a *saber butt*. The growth rings align with the dominant force (compression or tension), and the force is expressed in the formation of the rings. Schweingruber also notes, "Bent trees display the extreme dynamics of the regeneration process in crown and tree ring formation: growth reduction at the base of the trunk, compression wood develops on the lower stem side. The top bends upward, and the original dormant buds begin to sprout on the former stem branches. Newly formed, vertical shoots contain no compression wood" (8). Tree growth rings show the past struggles with the many forces nature creates.

The general growth rate of a healthy, normal-shaped tree can be an average of 6–10 growth rings per inch, depending greatly on the species of tree, the site conditions, and more. As a comparison, many of the recent growth rings we find on bent trees can have a range of 13–33 growth rings per inch.

False or Missing Growth Rings

False growth rings or frost rings are often triggered by a radical change in weather or climate within one season. A false spring is a common cause of false rings in North Texas due to a drastic change in weather over a short period of time. A false spring condition is created when an abnormally warm winter occurs, prompting trees to prematurely produce new growth (sometimes as early as February), followed by a severe freeze. An early freeze slows the biological functions of a tree, and it creates a latewood growth ring. As temperatures return to normal, the tree develops what appears to be another earlywood growth ring within the same year. Larson notes, "Frost rings are formed when the cambium resumes production of normal xylem and phloem. The degree of frost injury varies among species, the severity of the frost and its duration, and the time of occurrence. . . . In addition to formation of a frost ring resulting from cambial injury, severe frosts might also induce formation of a false growth ring (Ratzeburg 1869, 1871). As observed by Glock (1951), such rings occurred as extra rings following the frost ring in the radial increment" (1994, 575).

A tree reacts in the same fashion during an early-season drought, which slows its biological function but then eventually returns it to normal, creating the appearance of another growth ring. Extended droughts can cause the loss of one or more growth rings. Fire can also cause the loss of growth rings. Larson states, "Drought rings resemble frost rings in some respects, although cambial injury is usually less serious and recovery is more rapid. . . . Unlike the false rings resulting from frost, those resulting from drought are rarely preceded by either distorted or collapsed cells. . . . At the onset of drought, cambial cell division first subsides, followed by declines in cell expansion and maturation. At the expiration of the drought, cambial activity resumes and the only evidence of drought is the false growth ring" (1994, 575–76). Previous droughts in Texas were recorded by the Texas Water Resource Institute and published in the *Texas Water Journal.*

David William Stahl, PhD, of Arizona State University, published his research on false growth ring production and its relationship to false springs in *The Tree-Ring Record of False Spring in the Southcentral USA*. Stahl notes, "False spring conditions cause frost rings in oaks, and include both an abnormally warm winter and subsequent severe freeze in the spring" (1990, iii). His research covers false growth rings for a number of species, but those most pertinent to our current work include the post oak, *Quercus stellata*, and the Eastern red cedar, *Juniperous virginiana*. He studied growth rings from 1737 to 1982 on thirty-one sites. The Fort Worth site covered in his research is nearest to the Comanche Marker Trees discussed in this book. He counted 9,425 total rings, and 66 were frost rings, yielding a frost frequency percentage of 0.70.

According to Schweingruber:

> Enumeration becomes extremely problematic where false annual rings occur, or where the rings peter out or are missing entirely. It is impossible where the tree displays zones of irregular growth. In many cases the exact age of the tree can be determined only by applying the dendrochronological technique of cross-dating.
>
> Trees and shrubs in temperate and boreal regions generally form only one ring a year. The occurrence of clearly marked seasons in these regions means that very little variation in growth occurs within any given year.
>
> There can, by contrast, be enormous variations in growth in trees and shrubs in arid and semi-arid regions. This is a result of the unevenly distributed precipitation. A ring may be formed every year, or there may be no growth whatsoever, or the tree may even appear to form two or three rings a year. For samples in these regions, age can be determined only by cross-dating. (1998, 47)

Larson writes about the specifics regarding discontinuous growth rings and missing growth rings, "Actual missing rings are difficult to verify, and the only certain methods are comprehensive stem analysis, precise cross-dating, and comparison with trees of known age" (1994, 633).

Lost Limbs and Trees Provide Data

It is important to note that a tree can be lost or can lose a large limb after we have inspected it. Such events present an opportunity to learn more about its growth rate and potential age. As an example, a large bur oak in Bob Woodruff Park in Plano, Texas, was thought to be over two hundred years old. It is called Plano's "Bicentennial Bur Oak." This tree has a 60-inch DBH, it stands 90 feet tall, and it has an 80-foot-wide crown spread.

In 2006, a storm, accompanied by heavy winds, caused the loss of two large limbs at about 40 feet above the ground.

Fortunately, Plano city forester Renee Burke Jordan was alert and informed me of this unique opportunity. City forestry crews had made two large cuts on the tree, and there was a need to inspect the growth rings on the trunk. The tree cookie removed from the largest limb section that was

Bicentennial Bur Oak.

Photo of Steve Houser by Rex C. Curry, special contributor to the **Dallas Morning News**, *included in an article on the age of the Bicentennial Bur Oak by staff writer Jake Batsell, August 1, 2006.*

This tree cookie was preserved and turned on a wood lathe by Kevin Robert Bassett, creating a shallow bowl. Kevin is a highly talented woodworker, arborist, and plant pathologist.

not torn was 23 inches in diameter, excluding the bark, which was about 1.5 inches wide.

The 23-inch tree cookie was inspected by a number of experts, and growth ring estimates determined the approximate age of the branch to be between 215 and 226 years old. The growth rings are very close together, which leads to the conclusion that the tree is a very slow grower. Consider that it took many years to sprout the limb that fell from 40 feet above the ground, and the trunk DBH is 60 inches. If a 23-inch-diameter limb is 215 years old (on the low side), then the average growth rate per inch in diameter is 9.35 (215 divided by 23 inches). The average growth rate of 9.35 rings per inch times the trunk diameter of 60 inches equals 561 years old. Since normal upright-shaped trees grow much faster when they are younger, the tree is believed to be at least four hundred years old, depending on the expert consulted. By extension, a large limb lost from a tree, especially an Indian marker tree, must be analyzed because it can dramatically increase or decrease the estimation of the tree's potential age.

The big bur oak is the oldest living plant or tree in North Central Texas. It is an honor and a privilege to climb, care for, and be a steward for such a majestic old tree. At 400 years old, it has been a living witness to the presence and activities of American Indians for quite some time. Many generations have lived and died while this tree continues to endure all the harsh weather, insects, and pathogens that Texas can whip up. To be a tree climber gently rocking with the breeze in unison with the limbs of this tree restores the mind, body, and spirit.

If one of the trees we are researching is lost, an effort should be made to recover growth ring data for our files. Bent trees do not always grow like normal trees, and we need all the ring data we can gather on various species of trees, various shapes, and various environmental conditions. We encourage those who own a potential or recognized Comanche Marker Tree to keep us informed if the tree or a large limb is lost for any reason.

Sampling and Cross-Dating

It is important to try to record growth ring data for any tree previously removed near a potential Comanche Marker Tree. That means we always look for nearby stumps. Sampling more than just one tree on a site provides the growth rate of a normal-shaped tree in the immediate area and helps to determine if existing growth rate calculations are reasonable for a potential Comanche Marker Tree. The more ring data gathered on a site, the more accurate any projection becomes. In many urbanized conditions, however, nearby stumps are either not found or too much decay has occurred to read the growth rings.

With growth ring data from nearby trees, the rings can be compared through a process of *cross-dating*, which matches patterns in ring widths or other characteristics among a number of ring samples collected. Comparing growth ring samples from many trees in a particular location provides facts that lead to scientific conclusions. However, without a number of other trees that are bent in the same fashion in close proximity to the tree in question, any comparison of growth rings for cross-dating is not comparing "apples to apples." One is a bent tree while the others are normal-shaped trees. Experts conclude that without sufficient sampling and cross-dating, the results of dating the age of a tree may be unscientific. However, our options are limited, since core boring is out of the question, and nearby stumps may not exist.

Parameters Regarding Qualification

It does not take a large trunk radius to reach the requirement of 144 years of age, which potentially qualifies a tree as a Comanche Marker Tree. By comparing the radius of various tree sizes to the required age of 144, parameters can be established for the required minimum number of growth rings per inch using a simple formula: number of years to qualify divided by the radius equals the required minimum average number of growth rings.

- Example 1: 144 years divided by a 10-inch-radius tree requires at least 14.4 average growth rings per inch or greater.
- Example 2: 144 years divided by a 20-inch-radius tree would require at least 7.2 average growth rings per inch or greater.
- Example 3: 144 years divided by a 30-inch-radius tree would require at least 4.8 average growth rings per inch or greater.

We know that being bent when it is young will slow a tree's growth. How much and to what degree depends on the number of wounds created, the age of the tree when it was bent, the species, as well as many other cultural and environmental elements. Since average growth ring data are based on more recent growth in most cases, it is not reasonable to assume a tree would grow steadily at the average growth rate throughout its life. Calculating the minimum number of average growth rings determines if the growth rings found on a tree are small enough to potentially qualify. Many trees have been ruled out over the years due to a low growth ring count.

Based on a number of trees for which ring data were collected and analyzed over the years (sometimes on nearby stumps), a general pattern of growth for individual species in specific circumstances emerges. Our research involves only a few tree species, leaving us with a great deal of data regarding the average growth rate of a species under specific site conditions. As a result, the average growth rate of a tree can be altered by an *experience modifier* based on previous research. The radius of a tree is multiplied by the average growth rate, then adjusted by the experience modifying number. The modifier is adjusted as more research emerges in a specific part of Texas. As more tree cookies or stumps are analyzed, we gain a better understanding of the growth rate for a species in an area.

Growth rates can be reasonably deduced by examining many factors. The end result is an age range that is reasonable and based on the facts available at the time. It is not possible to project the exact age of a tree, but rather an age range based on the facts gathered to date.

If significant future evidence is uncovered that more clearly defines the age of a tree or alters other important aspects of our research, its status could change. We must base our research on what we currently know about a site or a tree, recognizing that there may be other evidence or research presented in the future that confirms or refutes our findings.

Our researchers keep in mind that the timeframe and age of a tree can be nullified by the Comanche Nation if significant information is presented to officials by its members. If a Comanche Nation member reports specific personal knowledge of a tree that could potentially affect its official status, the details of a tree are reviewed and its status altered, if necessary.

Determining the potential age range of a tree is an important first step in deciding if further research is warranted on any given tree. There are many bent or odd-shaped trees in nature, but without being old enough, they are nothing more than cool, interesting trees.

Maps of Recognized and Potential Comanche Marker Trees

STEVE HOUSER

CHAPTER 6

As potential Comanche Marker Trees are submitted for consideration, it often takes us time to sort through the details before we can determine if one or more trees have the real potential to qualify. Once a tree appears to be qualified, we give the site a number and create a corresponding electronic file. Sites with multiple trees are also provided a letter for each tree, which aids in identifying the trees individually. Site number 224 may have three trees, which are listed as 224a, 224b, and 224c in the files. Each tree is evaluated separately, based on its merit or the lack thereof.

After a file is created for a site, it often lies dormant for many months until volunteers travel to that particular part of the state. Once the site visit is complete, we develop a comprehensive inspection report, which includes the details about the tree, as well as research regarding the history of the site. One of the critical aspects of the research includes the development of various maps that help researchers draw conclusions or develop more hypotheses.

The GPS coordinates are recorded for each visited tree to allow its exact location to be plotted on a map. Individual sites with multiple trees are typically located on a map of the area to study any potential relationship between the trees. When a number of trees are located near each other in a general area, creating another map aids in understanding any potential relationship between them. In some cases, there does not appear to be a relationship. In others, the map helps our researchers find answers. For example, the direct connection between the California Crossing and the Bird's Fort Trail Comanche Marker Trees was recognized by studying maps and early trails (Each of these trees are discussed further in later chapters).

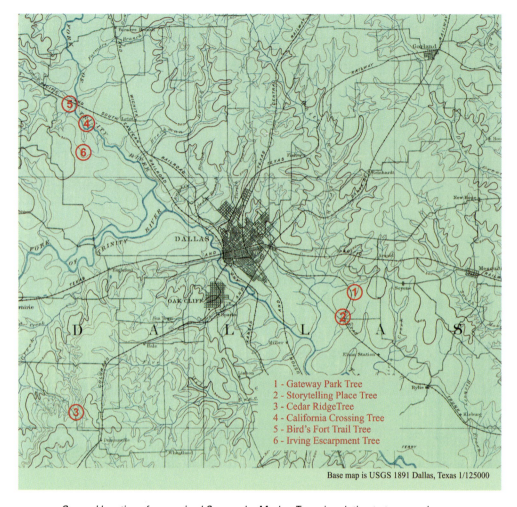

1 - Gateway Park Tree
2 - Storytelling Place Tree
3 - Cedar RidgeTree
4 - California Crossing Tree
5 - Bird's Fort Trail Tree
6 - Irving Escarpment Tree

Base map is USGS 1891 Dallas, Texas 1/125000

General location of recognized Comanche Marker Trees in relation to topography.

Texas Historic Tree Coalition trustee Jim Bagley stores the majority of our map collection. What follows are two maps he developed for the six recognized trees in Dallas, including their relationship to primary roads and water features. Three of the trees are located near the Trinity River.

Current technology gives the public access to many types of maps. Simply punching in an address on some Web sites (e.g., Google Earth or Bing) will provide an aerial view of a property. Often, our researchers review this type of map before we visit a site, to inspect the tree canopy cover in the

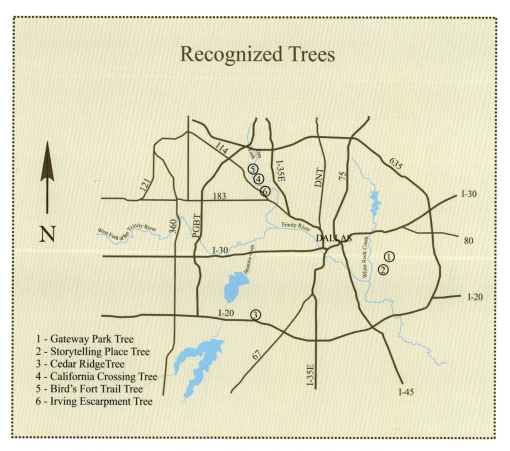

Recognized Trees

N

1 - Gateway Park Tree
2 - Storytelling Place Tree
3 - Cedar RidgeTree
4 - California Crossing Tree
5 - Bird's Fort Trail Tree
6 - Irving Escarpment Tree

General location of recognized Comanche Marker Trees in relation to highways.

area or to investigate natural features. Although the images are not high-resolution, they can help to determine the approximate location of a tree on a site before it is inspected. In the winter months, when trees are dormant, it is sometimes possible to find and view a larger bent-trunked tree on these sites. Our research in the area of mapping continues to evolve.

Our researchers have collected early trail maps over the years as important resources to help determine the previous movement of people through specific areas. We also study existing maps contained in various collections when we have the opportunity. Although they are not very precise in pinpointing the exact location of a trail, such maps can provide an approximate position that we can compare to the location of a potential Comanche

Marker Tree. One feature recorded about every tree visited is the direction of its point. Trees that appear near a trail and that point the same direction may have potential to be trail marker trees. If a number of trail marker trees were found along the general path of an old trail, the trees could help to define the exact location and route of the original trail.

Topography maps are an invaluable resource in determining the location of important natural features. We look very closely at existing and surrounding topography, as well as geologic formations on or near a site. It involves searching for features like a low-water crossing on a river or a high point used as a lookout or a smoke signaling area. These are natural features that would have attracted the Comanche. Understanding the landscape can help to determine the purpose of a tree or provide a theory regarding its existence.

Other maps that can provide valuable information include maps of existing water features. Rivers and creeks were an important source of water for the Comanche, but they were also often the location of valuable tree species that provided services (described in chapter 4, "Comanche Marker Tree Taxonomy") to the tribe. Bottomland areas often contained pecan, plum, walnut, and red cedar trees, among others. Rivers were a particular concern for travelers, and a low-water crossing often provided safe passage. Trails crossing larger rivers typically occurred at points with minimal embankment on the river and where the river spreads out wide with a shallow, rocky bottom.

Plotting trees on various types of maps requires time and effort, as well as expertise. Plotting the direction of point for multiple trees on a map adds another layer of complexity. During a visit with Comanche Nation Tribal Historic Preservation Officer Jimmy W. Arterberry in 2012, we provided detailed reports on eighty different sites containing trees that appeared to qualify. A discussion ensued regarding the need for mapping many potential Indian marker trees on various types of maps. Arterberry offered to contact Dr. Daniel J. Gelo to request assistance with maps that relate to our trees and natural resources. Dr. Gelo and his associates Thomas Hanson and Jason Roberts graciously offered their support.

This team developed various maps based on the GPS coordinates provided for each tree. A pointer was included to show the direction of point where applicable. Each tree was provided with the associated site number. It was important for us to develop a map showing all the potential Indian

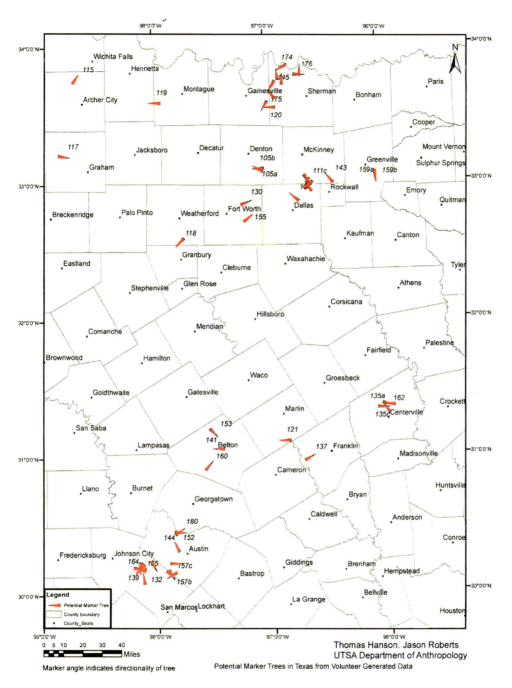

Comanche Marker Tree Project—Regional distribution of potential marker trees being researched and the direction they point.

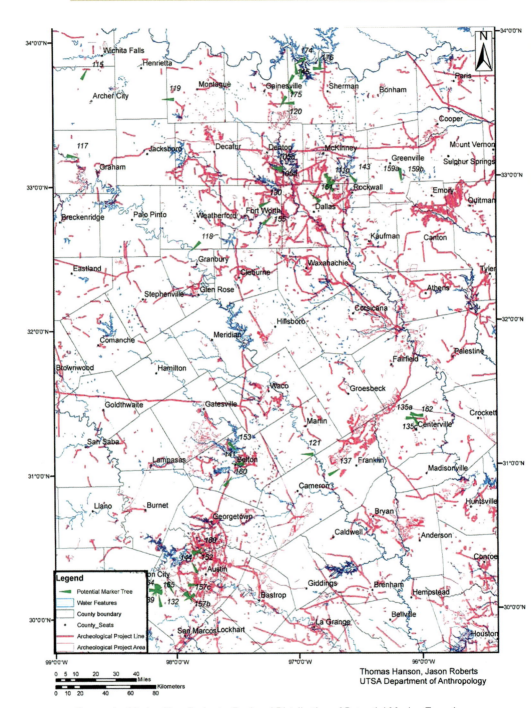

Comanche Marker Tree Project—Regional Distribution of Potential Marker Trees in Relation to THC Archaeological Projects.

marker trees and to look for patterns that might emerge, such as concentrations of trees or areas with no trees submitted.

Previous archaeological excavations and research can also provide clues about the past use of a particular site. A site that contained valuable resources was likely to be used by various civilizations in the past, and possibly by the Comanche in more recent years. A map showing potential Indian marker tree locations and their association with previous or ongoing archaeological projects provides a sense of history for the area and sometimes for an individual site. Finding a tree on a significant archaeological site related to the Comanche adds to the credibility of the tree and reinforces its potential to qualify as a tree marker.

It is difficult to view all the springs, creeks, or rivers on the previous map, but it is included to show what information this type of map *can* provide. When it is viewed on a computer with the ability to enlarge the image, the association between trees and water features becomes readily apparent.

Various types of maps can help to provide a great deal of information to support the qualification of a tree and may clarify the purpose assigned a tree, as well. However, we researchers remain aware that there are limits to the use of maps. Old trail maps are not very precise, and they do not provide a picture of an area's ecology as it existed many years ago. Recent water feature maps may show a lake that did not previously exist, or they may show the current route of a river or creek that traveled a different route in the past. Maps are a great resource, but they are only one of the tools that support the qualification of a tree.

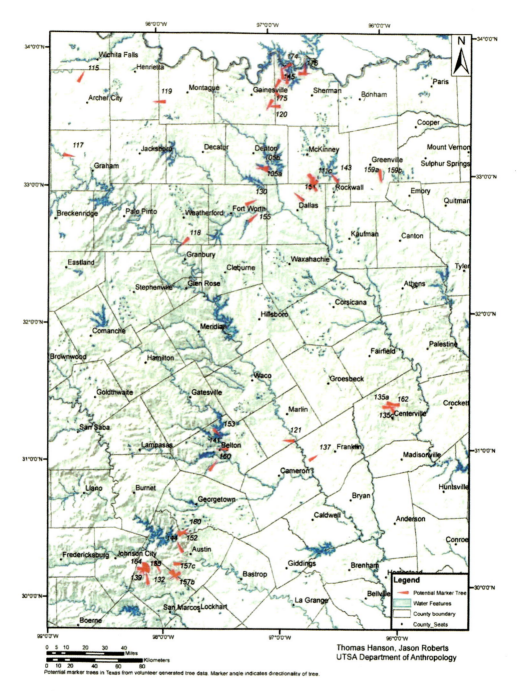

Potential marker trees in Texas from volunteer generated tree data. Marker angle indicates directionality of tree.

Thomas Hanson, Jason Roberts
UTSA Department of Anthropology

Comanche Marker Tree Project—Regional Distribution of Potential Marker Trees in Relation to Water Projects.

PART TWO

Tree Biology and Natural Forces

Nature Can Create
Bent Trees

STEVE HOUSER

T rees must endure many and varied types of stress. A tree in rural Texas must survive stresses such as severe drought or sizzling heat, and an inner city tree must also tolerate poor soil conditions or poor water quality. These types of stress can affect the health and reduce the life expectancy of a tree, but might also increase the likelihood that an entire tree will fall over (loss of support from the roots) or that limbs and trunks will suffer from structural failure.

All trees can fall or fail structurally, given the right conditions—often in ways that arborists and researchers do not fully understand. Reasons for failure can include anything from high winds to ice and snow buildup on the limbs to lightning strikes or uprooting. In a natural forested ecosystem, larger trees, limbs, or trunks are lost over time and fall on the understory of trees and growth. These actions can create trees that are similar in shape to those that are often claimed to be, or even determined to be, Indian marker trees.

Trees Can Adapt to External Stress

When a tree or a large limb falls on another tree, the accident can kill the injured tree or bend it in odd ways. The weight of the fallen limb or tree can crush the vascular system of a smaller tree to the point beyond which the smaller tree can function any longer. In cases where the injured tree survives, the tree reacts to the loading stress by adapting its growth and modifying its shape in response.

According to Gerhard Claus Mattheck in *Trees: The Mechanical Design*, "Adaptive growth can be defined as the elaborate shaping of individual tree

71

Snow accumulation on a small cedar elm, **Ulmus crassifolia.**

Similar cedar elm with excessive ice accumulation.

Red oaks, **Quercus shumardii,** *bent over and touching the ground due to heavy ice accumulation.*

designs well-adapted to external load conditions. No corrections or modifications are made unless external circumstances demand that they be made" (1991, 16). The "body language," or shape of an impacted tree, will often be seen for the remainder of its life. Mattheck notes that a tree's shape can reveal earlier loading situations that express the mechanical history of the tree. Certain shapes or growth forms can be identified immediately as consistent with a limb or tree trunk falling on top of another tree or limb. But in the cases of other shapes of trees our teams discover, it is often difficult to determine how the forms were created and whether they were created by humans or natural forces.

Imposters

The following photos of "imposters" were taken in East Texas eight years after the area had been clear-cut. Each of these trees appears to have the shape characteristics of an Indian marker tree, but they were undoubtedly shaped by natural forces and not intentionally bent by humans. Each tree is a native species and none is over 12 inches in diameter. If a large trunk falls on a small tree, it typically breaks or kills the tree. However, if a smaller limb from a falling tree lands on a smaller tree, it can create a permanent bend or a crook that will straighten out to varying degrees over time. A sharp bend that does not kill the tree will typically leave a permanent bend in the trunk.

This tree (opposite page), which has a 2-inch diameter, is growing out into a clearing, searching for sunlight. It will eventually grow more upright as the trunk becomes stronger, assuming it finds and gathers more sunlight on its foliage. If the needed sunlight is found in the direction it is currently growing (or at some other angle), the tree will continue to have a bow in the trunk. Researchers found this tree again at a later date on the edge of the clearing and discovered that its trunk had straightened to a more upright position, but still retained a slight angle toward the full sun.

This tree is about 7 inches in diameter and will likely retain the bend in the trunk for much or all of its life.

This tree is about 9 inches in diameter and will also retain this shape for many years to come.

Keep in mind that all these trees were found on one property. If this many trees were found on one site, there are likely a great number of trees bent in this manner around the state. Any sites that were previously clear-cut are quite likely to contain trees in these shapes.

Other imposters can be found in floodplains or low-lying areas. Trees in clear-cut areas are similar in shape to trees that fall on each other but flood waters can also create a similarly shaped tree. Previous flood or high-water conditions can cause a buildup of floating limbs and trunks that push over a small tree, forcing it to bend or bow. The wounds or scars on a tree will often indicate that it was once impacted by parts of another tree or external forces. Trees found in a park in Richardson, Texas, serve as examples of trees that are likely bent under such circumstances because they are not old enough to qualify as marker trees.

Although the clear-cutting of a forest by humans may have occurred

Small trees with bent trunks that may retain some of the bend over time.

Here is another tree in the clear-cut area that will likely keep its bent trunk for the rest of its life.

The long bow in the trunk of this tree will become much more upright over time as long as more sunlight is found above the tree.

An odd-shaped limb growing at an angle of almost 90 degrees from the main trunk. When a trunk or large limb pinned the end of this limb near the ground, it was not able to recover and grow at a normal angle.

This Eastern red cedar was toppled over by previous ice accumulation. The soil was raised about 6 inches on the side opposite the lean.

A different area of the site, clear-cut approximately thirty years ago, shows another imposter.

144 years ago, the practice certainly would not have been widespread. Since Indian marker trees must be a minimum age (144 years old or more, as of this writing) and it's unlikely that the practice of clear-cutting was wide-spread that many years ago, trees like these are very unlikely to have ever served as markers.

Another example of an imposter is an uprooted tree.

A small cedar elm, **Ulmus crassifolia,** *with a bow in the trunk.*

Another small cedar elm with a bend in the trunk .

A medium-sized green ash, **Fraxinus Pennsylvanica,** *with the shape of an Indian marker tree.*

Our research shows that the shapes of trees shown in these photographs are reasonably common. It is more uncommon, in walking through a forest, to find small trees that grow along the ground for quite a distance and then grow upward, or trees with a long bow in the trunk that touches the ground. Both of these shapes of trees are more difficult for natural forces to create.

If a small tree is bent into an arch from ice or snow, it will normally return in time to a much more upright position, assuming the vascular system is not severely damaged. If the weight of the ice is persistent for a longer period of time, a more permanent bend in the trunk can occur. Natural forces can indeed create trees that are imposters.

Even when nature created a bent or odd-shaped tree, the potential still exists for it to also be a Comanche Marker Tree. An odd or bent tree that stands out in the landscape could have been of use to the Comanche, regardless of how it was created. Based on what we discover about the

At a glance, this tree appears to be an Indian marker tree with three trunks.

Closer inspection reveals the fact that this osage orange or bois d'arc, **Maclura polmifera,**
was uprooted.

location, topography, and species of tree, among other factors, researchers can sometimes make the case that a tree created by nature was used by the Comanche in the past.

Pairing our other research findings with what we know about the unusual shapes of trees created by nature helps us separate the imposters from those with potential to qualify.

Bending Trunks and Limbs Alters Their Biological and Mechanical Functions

CHAPTER 8

STEVE HOUSER

U nderstanding the biology of a bent tree, and how it may differ from a tree with a normal shape, requires a basic knowledge of how a tree functions. Acknowledging that the subject is complex and involves an entire field of study, we offer here a basic explanation in layman's terms. We hope thoughtful readers will understand that a certain amount of accuracy can be lost through this oversimplification.

All above-ground parts of a tree have an outer covering of bark to provide protection and to hold moisture and nutrients inside the tree. Inside the outer layers of bark is a system of tubes or vessels known as *tracheids*, which carry water and other elements up and down the tree—think of it as a bunch of "straws" all bundled together. Each year, the tree grows new straws underneath its bark. The tree's roots pick up the water and nutrients, which then travel upward through the straws, or vascular system. These substances then are used by the foliage for the process of photosynthesis to produce the tree's food, which travels back down the tree through the straws. Water and elements are constantly moving up and down the tree as a part of its normal biological process.

A severe bend or additional stress added to the straws of a tree can affect its ability to function normally. To create a Comanche Marker Tree, or any bent tree, required instilling the bend in the trunk when the tree was small. Bending methods included tying the tree down with a thong made of animal hide or yucca rope, or weighing the tree down with rocks tied together, forcing it to bend.

When a bend is forced in a tree's trunk, the biological process can be slowed to a great degree, depending on the size of the tree and its species, among other environmental and cultural conditions. The growth of the tree can be stunted throughout most of its life, or in some cases, the tree's biology can adapt to the changes. How much and to what degree depends greatly on the severity of the bend, as well as the species and various other cultural issues. In some cases, trees never seem to grow normally again, while in others, they seem to adapt reasonably well to the bend.

Potential Sizes Used

We know that native trees growing in the right locations were altered for specific purposes. If any tree was bent too far in the beginning of its life, its trunk could break. The tree could also die from a lack of moisture and nourishment caused by the severe bend in its vascular system. As a result, the process of creating an Indian marker tree may have involved slowly bending the tree a little at a time and may have included the removal of wood on the outside of the bend to relieve the stress.

Bending the trunk of a tree at a severe angle causes a compression of the straws on the underside of the bend and stretches the straws along the top of the bend. As a result, the bent trees we find today often have bark that still appears compressed on the underside, and many show old wounds or scars on the top side. The wounds on the top can be attributed to the force of bending or to wounds intentionally created to relieve the tension caused by being stretched.

Various sizes of trees could have been altered, with each size having benefits and drawbacks.

Tree with a 1- or 2-inch diameter:
- A tree this size would likely be too small to suit the intended purpose of a marker tree until it grows for a number of years. Being perhaps only 1 or 2 feet in height before it was bent, it could easily be hidden by surrounding plants, making it hard to find again in the future.
- The vascular system (or straws) would be able to adapt to the bend more readily than a larger tree.

Tree with a 2- to 3-inch diameter:
- Being 4 to 6 feet in height, the tree would possess improved visibility at a distance and would be easier to find.

Compressed bark on the underside of a bend.

- The trunk would still be young enough for the wood to be pliable and bendable.
- A little more pressure would be required to bend the trunk, as opposed to the pressure required by a 1- or 2-inch tree.
- The tree might require wounding to relieve the stress from bending.
- The vascular system could potentially adapt and recover within a few years.

Tree with a 4- to 6-inch diameter:
- Possibly 8 to 10 feet in height, the tree would have improved visibility at a greater distance and be easier to find.
- The tree would resist being forced into a sharper angle in the lower sections without a greater amount of pressure.
- The tree would require wounding to relieve the stress from bending.
- The vascular system might recover, but the healing process could take years.

To create a severe bend in the lower section of a tree that is larger than 6 inches in diameter might be possible, but difficult due to the amount of pressure required on the trunk and the attendant damage to the vascular system. The tree would likely break or die due to the inflicted damage. Bending a tree larger than 6 inches would require a considerable amount of effort and time, as well as a strong knowledge of the individual characteristics of the tree species. As a result, trees less than 6 inches in diameter appear to have been the size most likely used in the past.

A number of trees currently under investigation show old wounds on the top side of the bend. Trees must produce callous tissue to heal or compartmentalize an earlier wound, often leaving a scar for the life of the tree. It is possible that trees with wounds on the bend of the trunk were larger trees when they were originally bent.

Impacts of Bending Trees

The larger a tree was when it was bent, the greater the potential impact on the current and future health of the tree. Since the up and down movement of nourishing materials is critical to the health of a tree, pinching or stretching the straws is much like pinching or bending a garden hose. The outer section of the bend in the hose would be greatly stretched, causing other parts of the hose to shrink in size. If you put a sharp bend in a hose, it pinches the straw, which reduces or eliminates the flow of water.

Bird's Fort Trail Tree.

Trunk of the Bird's Fort Trail Tree containing a long scar from an old wound.

Trunk scars from old wounds.

Trunk with numerous cross-grain scars.

The compression and stretching can have a profound effect on the growth rate and health of a tree almost immediately, especially if the bend results in the foliage being shaded by surrounding trees. Most species of shade trees need full sun to stay healthy and grow at a normal rate. Without full sun they will not produce as much food, and their biological functions will slow. The result is slower growth and smaller growth rings (or straws) each year.

A tree's physical response to being severely bent can depend not only on the size of the tree but also on the strength and pliability of the wood, which can vary greatly by species. Soft-wooded species of trees are more pliable and easier to bend, but they are also more apt to break when the tree grows larger. Soft-wooded tree species could be successfully bent when they are larger in size, as opposed to harder-wooded tree species. As a general rule, smaller or softer-wooded tree species adapted better to the bend than did larger or harder-wooded trees.

Phototropism and Geotropism

Since the American Indians used a yucca rope or a thong and possibly rocks to bend over a small tree, it would have taken at least a year or two of being held in place (depending on the species and any damage done during bending) to force a bend in a tree that would not change after being untied or released. Even if the vascular system adjusts to the alteration, there are two forces that will always affect tree growth or regrowth after bending: light and gravity.

Since the direction of a tree's growth (or regrowth) is directly influenced by light and gravity, removing a thong and rocks before the vascular system adjusts to the bend will allow the tree to grow back into an upright position. If the vascular system has not adjusted, the tree will tend to revert back to a normal shape. The need for a tree to always grow toward the highest concentration of sunlight is called *phototropism*. As an example, a small tree growing at the edge of a grove of larger trees will typically grow out at an angle (away from nearby trees) to garner as much sun as possible.

Trees that have fallen over but still survived will always try to grow back to an upward position. The need for a tree to always grow upright (or against gravity) is called *geotropism*. So even though a trunk may be bent, it will tend to grow upright after the bend, assuming there is full sun above it.

When a small tree is bent and mostly full sun bears down on the trunk, most tree species will have a tendency to produce upright sprouts along the trunk in order to shade the bark from the full sun. The proper term for the upright growth is *epicormic sprouts*, but most people refer to them as "suckers." These sprouts are generated in response to sunlight but are influenced by geotropism, typically growing straight up or toward full sun.

Tree Species and Longevity

The tree species often used for Comanche Marker Trees include pecan (*Carya illinoensis*), post oak (*Quercus stellata*), bur oak (*Quercus macrocarpa*), and Eastern red cedar (*Juniperus virginiana*). Other tree species have been submitted to us and may qualify for recognition in the future. The Comanche clearly had a great deal of knowledge about various tree species, including which trees would live a long life. It takes special skills and training to know how to form a tree into a specific shape for a specific purpose. This knowledge was passed down orally through many, many generations because of the importance of these trees to the Comanche culture.

We know that the longer-lived tree species are those most likely to be found today because the shorter-lived species are less likely to have survived the requisite 144 years. However, there appear to be some exceptions to the rule.

One of the trees currently under investigation by researchers was submitted by Mr. George Blackmon of Bowie, Texas. Mr. Blackmon submitted a potential Comanche Marker Tree that was thought to be a bois d'arc or osage orange (*Maclura pomifera*). It turned out to be a red mulberry (*Morus rubra*), a species that typically lives to be around seventy-five years old. This is not a typical tree, however, and we often call it the "pretzel" tree. It is a single-trunk tree with multiple limbs growing in various directions and has three different root systems.

The second photo shows Mr. Blackmon and his grandchildren playing on the pretzel tree. Since the tree was bearing fruit at the time, the grandchildren must have discovered that the mulberries were ripe. In the photo, they are shown scampered through the tree picking fruit, making it clear why fruit-bearing trees were kept low to the ground by the Comanche. Even the children could reach the fruit. The Blackmon grandchildren are shown giggling as they practiced their climbing skills and were soon covered in red mulberry juice. Children love Indian marker trees for more than one reason.

The trunk of this unusual tree grows out of the ground on the right side of the photo and touches the ground where the girl is sitting. It roots into the ground again in this area, creating a second root system. There is an upright bow to the limb where Mr. Blackmon is sitting, and it roots into the ground a third time on the far left side of the photo.

At several feet above the ground, the main trunk is greatly enlarged or bloated. This type of enlargement is not unexpected in marker trees. Trunk enlargement occurs when something binds a trunk for a long period of time. When the straws are pinched to the point of restricting or cutting off the flow of elements and water, the back-pressure forces the straws to become enlarged. As the trunk of the pretzel tree emerges from the ground at an angle, it is round for the first few feet. At this point, it becomes bloated, as if an external force severely bound or girdled the trunk for an extended

Mr. Blackmon and his grandchildren.

A closer look at the pretzel tree.

period of time. Any materials used to tie down a tree that circled much of the trunk and were not removed or biodegraded within a few years could have left a permanent enlargement in the vascular system.

Beyond the bloated trunk, the tree touches the ground and then splits in opposite directions. One stem is bowed across the main trunk just before the enlargement. The other stem bows the opposite direction, producing a pretzel-like shape.

As our team's research progresses, we hope to discover for certain whether or not Mr. Blackmon's tree is a Comanche Marker Tree.

Knowing how trunks and limbs can be manipulated by natural or human forces provides us some of the information we need to determine a tree's previous history. Understanding the body language of a tree helps to tell its life story.

Roots growing off the trunk where it touched the ground on the left of the previous photo.

Roots growing off the trunk under the girl in the previous photo.

The enlarged area also contains a distinct bend.

The opposite side of the trunk is enlarged, as well.

One trunk contains a severe bend, causing it to rest on the main trunk.

Alterations by external forces can create a very unique shape of tree.

PART THREE

Profiles of Comanche Marker Trees

Gateway Park Comanche Marker Tree

Lower White Rock Creek Section of Dallas's Great Trinity Forest

STEVE HOUSER AND LINDA PELON

Background

BY LINDA PELON

My research on Indian marker trees began while discussing thesis research on Comanche ethnohistory and land use with a neighbor, Rose Flood, and her friend, Jim Nichols. I mentioned being curious about these bent trees that some ranchers were reporting were "Indian marker trees," though archaeologists I talked with dismissed them as "ranch lore" and authoritatively asserted Indian marker trees did not exist. Both Rose and Jim, senior citizens who had spent their earlier years in West Texas, stated that they were aware of these bent Indian marker trees, and Rose informed me that there was one that often made her wonder whether or not it was an Indian marker tree, located in a park within blocks of our neighborhood. She took me to see it, and this began a long journey that I fondly call "the research that won't go away."

Soon after visiting this tree, documentation was found of a Comanche Marker Tree in a book by the Texas Forest Service, *Famous Trees of Texas*. Since the focus of research at the time was on a Central Texas Comanche band, the Penatuhkah, I was under the impression that Comanche land use did not extend as far east as the Trinity River. I decided to do some research on the area around this bent tree in Gateway Park but did not expect to find information linking this tree to Comanche occupation of the area. Arborists Steve Houser and Bill Seaman as well as state forester Courtney Blevins were consulted to determine a potential age range for the tree. These professionals

99

agreed that the tree was at least 150 years old; it could not be ruled out as a Comanche Marker Tree based on age. In addition, they found no evidence indicating that this tree had been unusually shaped by natural causes.

The next step was to survey the historic record for information linking this area to a Comanche presence. It was soon discovered that surveyor Warren Ferris was forced to dodge Comanche war parties while surveying Dallas County in the 1840s. Earlier research indicated that Comanche tradition is to defend land that they claim. These war parties were not just passing through but were chasing the hated surveyor. Comanche people recognized survey equipment and its purpose. Their word for this equipment translates as "the things that steal the land." I also discovered that this area of the Trinity River valley was once a gathering place for trade between the agricultural Indians and the plains hunters. Dallas is now what it has always been—a multicultural trading center.

An intriguing reference to smoke signaling was found in a book on the early history of Mesquite, Texas, that included the southeast section of Dallas's Trinity River corridor where Gateway Park is located. Published by the Mesquite Historical Committee, *A Stake in the Prairie* includes this reference to Indians in the area in its opening paragraph: "Moving up from the east on horseback came the all-powerful *Caddi*, followed by his retinue of Caddo chieftains known as *canahas*. Mounting a ridge of land overlooking the river bottom, the group looked westward to the high rise of cedar hills far across the river's forks for signals of the leaders from western tribes who were now also journeying to the trade fair" (1984, 7). A Comanche smoke signaling system was well known to army officers and others traveling in areas of Texas occupied by Comanche. A southwest-facing ridge within a few blocks of Gateway Park provided a strategic scenic view from which other high points to the west can be seen. Could that nearby high point be the easternmost signaling point in the Comanche signaling system?

It was soon evident that this area fit the template for a preferred Comanche campsite. Gateway Park is on Oak Creek, which drains into Lower White Rock Creek near its confluence with the Trinity River. Oak Creek is lined with groves of native pecan trees, and many freshwater springs occur along its path. The adjacent escarpment ridges include some of the highest points in Dallas County. Springs for fresh water, high points for surveillance and smoke signaling, and pecan-lined creeks near major rivers were features present in many large Comanche campsites. And the pecan nuts, referred to as "protein that won't run away" by Texas Tech archaeologist

Grant Hall, were an important part of the Comanche diet. In addition, nearby Scyene Road was described by a Texas historical marker as having once been a buffalo trail. A survey completed by the Dallas Archaeological Society (DAS) documented numerous archaeological sites in the area, indicating heavy camping by Indians.

All this research about the bent tree in Gateway Park was organized into a written nomination submitted to the Comanche Nation, and in 1997 the Comanche Nation proclaimed the tree a Comanche Marker Tree. This tree lived for about two more years before it was destroyed in a Memorial Day thunderstorm. Research continues to document additional landscape features significant to Comanche cultural heritage in this area. Archaeological evidence of a large woodworking worksite was unearthed and destroyed by a land developer, but artifacts from this site and its location were documented. The Comanche Language and Culture Preservation Committee visited the area around the adjacent Scyene Overlook (the suspected smoke signaling station) and found that many of the plants traditionally used for medicinal purposes still grow in the surviving prairie grass meadows below the overlook. These medicine plants, combined with a spring-fed flowing stream and a high point for prayer and meditation, are key features for identifying a Comanche Medicine Place.

Finally, in a subsequent visit, Jimmy W. Arterberry, Comanche Nation tribal historic preservation officer, identified a Comanche Storytelling Place on a nearby highpoint south of the Scyene Overlook, as well as "Lodge Pole Woods" growing along the base of the escarpment ridge. These tall, straight red cedars were a preferred building material for Comanche lodges. Work is in progress to follow through with a recommendation by Comanche leaders to nominate this entire section of Dallas's Great Trinity Forest parkland along Lower White Rock Creek to the National Register as a National Historic Landmark District with Traditional Cultural Properties. This will provide recognition for all the natural and cultural features associated with the Gateway Park Marker Tree and preserve this area as an archive of cultural and natural resources.

Arborist Perspective

BY STEVE HOUSER

The Gateway Park Comanche Marker Tree was a pecan (*Carya illinoensis*) located in Gateway Park (East Dallas). The park is near the eastern edge of the Great Trinity Forest, reported to be the largest urban bottomland forest in the country at almost 8,000 acres. By comparison, New York's Central Park is 6,160 acres.

In 1959, before the importance of the tree was recognized, a new neighborhood was established near the Gateway Park Comanche Marker Tree. Fortunately, the tree was located in the designated parkland and was not removed as a result of the development. Unfortunately, as development continued, a power line was installed near the tree and most of the top portion of the tree was removed to accommodate the utility lines. At some point in the past, limbs that emanated from the trunk where it touched the ground and upward for a number of feet, were removed. The result was an extensive amount of decay in the trunk from the ground level upward.

Old cut wounds and extensive decay extended from the ground level upward to around 4 feet from where the trunk touched the ground. Decay started in this area and eventually spread throughout most of the trunk. Other wounds were evident in many parts of the trunk, due to some type of physical injury, with the most likely suspect being damage from previous human activity. The wounds were not just superficial in nature; they extended into the

Staff photo courtesy of Summer Zak that appeared in a **Dallas Morning News** *article, Sunday, April 27, 1997, by Berta Delgado and appeared again in a second* **Dallas Morning News** *article, Saturday, December 24, 2004, by Larry Bleiberg.*

inner portions of the tree. These wounds effectively stopped the flow of water and essential elements through this section to the top half of the tree. The tree was in poor and declining health due to the removal of top growth done years ago, the extensive amount of decay, and previous physical damage to the trunk. Additionally, the bowed shape of the trunk made it difficult for the tree to carry out its normal biological functions. Water and nutrients must travel a long distance through a long, bowed trunk, which likely would have affected the health of any tree with this shape.

Formal Recognition

The tree was formally recognized in April of 1997 by proclamation from the Comanche Nation. The event was documented in a *Dallas Morning News* article by staff writer Berta Delgado on Sunday, April 27, 1997. A follow-up article by Rebecca Perry, a special contributor, was printed in the *Dallas Morning News* on November 11, 1997. The official proclamation from the Comanche Nation was of historic significance in a number of ways. First and foremost, it reaffirmed that the Comanche Nation once inhabited the Great Trinity Forest area. While it was surprising to many historians to find a tree fitting the description of a Comanche Marker Tree as far east as the Trinity River, knowledgeable Comanche individuals were not surprised. They had not forgotten about the importance of *Pih-heet Pah-e-hoona*—the Comanche name for the Trinity River; the term translates as "Three Rivers."

The area in the vicinity of the tree was explored from a Comanche perspective and found to contain a treasure trove of resources important to the lifestyle of these nomadic people. In fact, the site fits a template for a preferred Comanche campsite. Archaeological surveys previously conducted by the DAS documented extensive use of this area for camping by native people. One DAS officer commented that the area was "so saturated with sites as to be considered one big campsite."

Second, the occasion marked the first time ever that a Comanche Marker Tree was officially recognized by the tribal government. The proclamation, signed on April 7, 1997, by Chairman Wallace Coffey, recognized the Gateway Park Marker Tree as "a living monument to our historic presence in the great state of Texas" and noted "the importance of this tree to Native American cultural heritage."

Comanche poet and educator Juanita Pahdopony was sent as an ambassador from the Comanche Nation to read this proclamation at the dedication ceremony for the tree on April 26, 1997. She was quoted in a report in the *Dallas Morning News* (April 27, 1997): "This represents a continuation of our culture. We are one with nature. This tree has a lot of stories to tell. It is too bad that we have not been here to hear them."

Third, after 144 years of exile and disconnection from their Texas homelands, sacred places, and heritage sites, the Comanche Nation finally reconnected with the *Pih-heet Pah-e-hoona*. The preservation of these natural features and heritage sites, and the tranquil natural setting in which they exist, is of great concern to the Comanche Nation as well as to many Texans who also value the diversity of heritages and histories in Texas.

The recognition event included anthropologists such as Linda Pelon, who organized the event, and it was hosted by the Dallas Historic Tree Coalition along with the Piedmont-Scyene and the Parkdale Heights neighborhoods. Dallas Historic Tree Coalition President Mary Ellen Bluntzer noted, "This is the way human nature should be, coming together to recognize something that can never be replaced." Comanche Representative James Yellowfish said a prayer, and the group sang Comanche songs.

Comanche Blessing

Within a few months of the dedication event, Chairman Coffey visited the site and completed traditional blessing ceremonies for the area from the scenic overlook on the escarpment ridge above the Keeton Park Golf Course.

Because the tree appeared damaged and unhealthy, Glenn Watson, pipe carrier for the Comanche War Scouts, visited the site often to pray for the tree in the traditional Comanche way. The Comanche Marker Tree was discussed often on *Beyond Bows and Arrows*, the community radio program serving Dallas's Indian community, resulting in many additional visits to the tree. Local Indian families brought their children to see it, and many included the tree in their prayers.

Tree Offspring

A few months after the dedication event, Dallas Historic Tree Coalition officials and other interested parties worked to collect seeds (pecans) from the tree. They were mailed to the nonprofit organization American Forests to be grown in their historic tree nursery in Jacksonville, Florida. At the time, the American Forests group sold seedlings from trees connected to historic events, places, and personalities. For a number of years, the off-spring were sold around the nation as historic seedlings.

In the fall of 2004, Hurricane Frances destroyed all the remaining saplings from the Gateway Park Comanche Marker Tree being grown at the American Forests nursery in Florida. The nursery had already sold around 150 of the offspring when the remaining trees were lost, so it is possible that some of the original trees planted may be producing pecan seeds today. After the offspring were growing quite well, we learned that Texas agricultural laws do not allow pecan seedlings or saplings to be mailed or shipped into the state, due to concerns over the transportation of insects or diseases.

Tree Loss

Unfortunately, within two years of the official recognition, high winds accompanying a Memorial Day storm (1998) caused the entire top of the Gateway Park Comanche Marker Tree to break at the soil level, leaving only a long bowed trunk. The loss was a devastating blow to those who recognized the tree's full value to society and to the history of the Comanche culture. It was also ironic that the tree was lost on Memorial Day; the tree lasted long enough to provide offspring, and the day it was lost seemed to commemorate those who created and cared for it over many years. The loss was featured in the *Dallas Morning News* on Sunday, March 26, 2000, in an article by staff writer Louise Applebome.

Revival Attempt

Since foliage produces food for a tree through photosynthesis, those concerned recognized that the loss of all the foliage might cause the death of what remained of the tree. Pecans are tough trees and often regenerate new growth after an extensive loss of foliage. As a result, we waited through the early part of the next growing season to determine if the tree had enough energy to regrow sprouts, or leafy growth, along the trunk. When this did not occur, there were deep concerns, and an "emergency surgery" was performed as a last resort. In discussions with two plant pathologists, Russell Peters and Kevin Bassett (both dear friends of the coalition), we concluded that we might try to plant small pecan saplings near the base of

*Staff photo courtesy of Chris Hamilton from a **Dallas Morning News** article, Sunday, March 26, 2000, by Louise Applebome*

the tree and graft them into the trunk. This would provide an "IV" for the tree by supplying nutrients for the straws in the trunk, since there was no top growth or foliage. It was a long shot because of the poor condition of the trunk, which may not have been healthy enough to accept the graft, but the small saplings might have provided enough energy to encourage the trunk to resprout some growth (or epicormic sprouts). The operation and IV ultimately failed, but some comfort was gained by knowing we had made an honest effort to revive the tree.

Tree Cookie

On a positive note, a tree cookie, or thin slice of trunk, was recovered from the historic tree. Since the purpose of removing a tree cookie is to count and document growth rings, a solid piece of wood is required that is not missing many growth rings in the center. For that reason, a tree cookie was taken just above the area of extensive decay over the point that the trunk touched the ground. This turned out to be a good decision because the remaining arched trunk was removed and the stump was ground out before any concerned parties were aware the removal was scheduled.

The last known photographs of the arched trunk were taken by an arborist with a Polaroid camera. Scans of those prints are reproduced here.

Dating the Age Range

The tree cookie was removed at 14 feet above the ground. The basic measurements are shown in the following rough sketch:

The 13-inch section was sanded to aid in counting the growth rings and it was treated to ensure it would not dry out and split. The rings were then counted many times from various angles, counting only light or dark rings. (Trees produce both spring and summer wood in each growing season, one with darker growth rings and one lighter in color.)

The size of the tree cookie appears to be slightly less than 13 inches because the bark was lost in the process of sanding and treating the cookie. After checking with a number of highly experienced arborists who also counted the rings, it was concluded that the cookie was 118 years old at the time it was removed, meaning that that portion of the trunk dated back to 1892.

▲ Long arching bow in the trunk.

▲▶ Arching trunk with trees in the background.

▶ Area containing decay where the trunk touched the ground.

If the 13-inch tree cookie is 118 years old, how old is the 64-inch base? Trees typically grow faster when they are young. How much depends on a number of factors, such as the species and soil conditions. Since the base of the tree and the trunk were both hollow, it is likely the tree was damaged in the past, causing decay to spread in the base and trunk. Bending a sapling at severe angles and strapping yucca rope around the trunk, as well as removing wood to relieve the stress, would greatly slow the growth of a tree in its early years.

A close study of the tree cookie growth rings shows either previous droughts that occurred in the area or a loss of foliage at the time, which slowed the growth. As an example, the growth rate slowed greatly around the time that the top was removed for the power line installation in 1959. This is judged by looking at the growth rings around this period of time. Other examples include irregular growth near the center, followed by suppressed growth for ten years, early in the life of the tree.

Growth suppression can occur for a number of reasons. For trees unaffected by humans, drought is the most common natural cause. A study of past droughts and weather will often correlate directly to the growth rings we find on trees in an area. Extensive ice, snow, or wind damage to foliage and the subsequent loss of limbs can also slow the growth rate and create tighter growth rings.

For trees affected by human activities, such as root damage from trenching or excessive removal of foliage, the growth is often suppressed. The loss

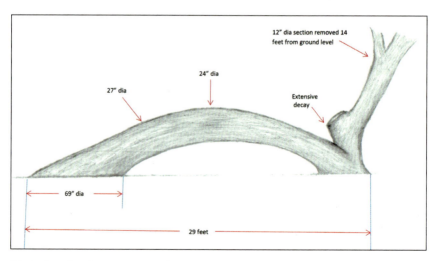

Illustration showing measurements.

Gateway tree cookie with ruler.

of roots creates a loss of moisture and other elements that the tree needs to function and stay healthy. An excessive loss of foliage (or more than 25 percent) may suppress the growth rate because foliage produces food for the tree through photosynthesis. The loss of a significant amount of foliage reduces the amount of food (sugars and carbohydrates) a tree can produce, with much the same results as cutting a person's diet. Likewise, the loss of roots can reduce a tree's ability to transfer the elements needed to produce food.

Although the approximate age of the historic tree could be argued, available evidence indicates the pecan was fully old enough to qualify as a Comanche Marker Tree.

As the first Comanche Marker Tree ever recognized, the Gateway Park tree opened the eyes of many to the fact that these trees still exist today. The recognition marked the beginning of our research and a better understanding of a subject about which little was previously known in Texas.

▲ *The Storytelling Place tree in dormancy. Not all marker trees are bent.*

▶ *The Storytelling Place tree during the growing season.*

Storytelling Place
Comanche Marker Tree

STEVE HOUSER

CHAPTER 10

Arborist Perspective

The Storytelling Place Comanche Marker Tree in East Dallas is located at the top of a natural ridge that surrounds the Great Trinity Forest, reported to be the largest urban bottomland hardwood forest in the country. There, at the top of a ridge, the soil composition is mostly white limestone rock with very little other soil near the surface. The rock is fractured limestone and chalky in nature, and the other existing soil is black clay with moderate alkalinity. The tree itself sits on solid rock with very limited topsoil. The site is east of the native Cross Timbers ecological area, which includes soils that are sandier and more acidic in nature.

The tree is a Texas red oak (*Quercus texana buckleyi*), which is a shorter-growing species that seldom reaches more than 50 feet tall and has a broad, spreading crown. These red oaks are often multistemmed trees with leaves similar to those of the Shumard red oak (*Quercus shumardii*). The Texas red oak is a much more drought-tolerant species than the Shumard red oak and typically grows at a much slower rate. The Shumard oak red prefers the richer soils found in bottomland areas, whereas the Texas red oak will tolerate rocky and drier soils, which are lower in nutrient content.

Although there are many larger trees in bottomland areas of the forest, this particular tree was always considered one of the older trees in the watershed by environmental pioneer and longtime Dallas resident Ned Fritz. Fritz often led nature tours of the Great Trinity Forest to educate, inspire, and encourage others to care for the forest and all its inhabitants. These missions are currently carried out by members of the North Texas chapter of the Texas Master Naturalist Program and the Trinity River Audubon Center.

White rock exposed on the surface limits annual tree growth.

A mix of Texas red oak and Eastern red cedar found along the ridge.

This particular ridge of limestone is well known for its abundance of Texas red oaks, Eastern red cedars, post oaks, and many other drought-tolerant plants and trees.

The tree is not large. Its height is 28 feet, with a crown spread of 40 feet. There are seven trunks ranging in size from 6.5 to 12 inches in diameter. The base of the tree is approximately 56 inches in diameter.

Although the species is prone to having multiple trunks, each one is not growing out of the ground but rather out of the remnants or stump of the tree, lost many years ago.

What caused the loss of the original tree is not known, but it is likely to have been natural causes, such as severe drought, significant ice buildup (that caused limb breakage), or high winds. The loss of all top growth is referred to as decapitation because most or all of the top foliage is lost. According to Philip R. Larson in *The Vascular Cambium*, "Removal of a growing bud or shoot by decapitation is usually followed by a subordinate axillary bud growing out to assume the role of a new terminal" (1994, 544). In the case of this tree, multiple buds were regenerated, which led to numerous limbs growing off the original stump over time.

Multiple trunks growing from a single base.

Multiple areas of decay from the original trunks lost in the past.

Growing on the top of a ridge, the tree receives no runoff, and the limited top soil on the site would only hold a minor amount of water for a short period of time during a drought. There is limited tree cover close by, leaving much of the root system open to full sun, which would increase the evaporation of moisture from the soil. The more sun there is on the roots, the less the tree can hold and retain rainwater. There is also very limited to no organic matter on or near the soil surface because the winds blow the leaves and smaller twigs downhill or away from the tree. Organic matter (leaves and twigs, or mulch) helps to hold soil moisture and provide many other benefits including the reduction of soil evaporation due to the shade it offers to the soil.

As further evidence that the original tree was lost, another Texas red oak on the same ridge at the same level and 40 yards to the north, shows the exact same regrowth around the base of a stump.

Both trees show a significant amount of soil erosion, which exposes the roots near their bases. It is not clear how many of the exposed roots were caused by erosion or by the trees' producing new roots—often called

Sprout growth on a stump eventually turns into trunks.

Roots appear to be several inches above the ground level due to erosion.

buttress roots—off a trunk. Such buttress roots can help the tree to absorb more water and elements. If the exposure of roots was due primarily to erosion, the loss of about 4 inches of soil would have taken many years.

As a force of nature (e.g., ice, snow, or wind) acts to remove much of the above-ground portions of the tree, the roots often remain healthy and active. As a result, the tree often reacts to the damage by regrowing multiple stems either from the stump or from the remaining healthy underground roots.

The trunks of trees in this area often exhibit regrowth after the loss of an original tree. As a result, dating the age becomes more difficult. The growth rings from samples taken were rings from the regrowth rather than the growth of the original trunks, previously lost.

The base of the Comanche Storytelling Place Tree is approximately 56 inches in diameter, depending on where it is measured. Close examination of the base indicates that the original tree was not a single-trunked tree. It was more likely to have been two original trunks, which were lost many years ago. The ruler in the following photo shows the location of a decayed area from one of the original trunks.

Multiple trunks growing from the base of a lost red oak nearby.

Larger area of decay from an original trunk.

Exposed roots due to soil erosion.

Dating the Age Range

Dead limbs removed from the tree indicated the recent growth rate to be a wide range of 18–27 growth rings per inch in diameter. Due to poor soil conditions, the growth rate was very slow. During previous drought years, growth rings were difficult to discern due to their close proximity.

Since the original trunks of the tree are either covered with bark or severely decayed, it is not possible to read any growth rings. The exact size of the original trunks is not clear. If there were two original trunks, the size of each trunk could have been as large as 28 inches or half the size of the tree at the base.

Since the original tree likely included two trunks of 28 inches, the average growth rate required to reach 144 years old is 10.28 growth rings per

Holes in a tree cookie are often created by wood-boring insects.

Tighter growth rings often require a magnifying glass to read them.

inch (144 divided by the radius of 14 inches for each trunk). However, any age calculations must include the trunks that regenerated after the original tree was lost. The current growth rate of the regenerated trunks varied from 18 to 27 growth rings per inch or much more than the average number of growth rings required (10.8). Although the approximate age could be argued, based on current growth ring evidence, the original tree is old enough to qualify as a Comanche Marker Tree.

Normal-Shaped Trees Can Be Markers

To quote Comanche Nation Historic Preservation Officer Jimmy W. Arterberry, "All trees can be marker trees, but are not." This tree is an example of one that is not bent nor is it a famous "Treaty Tree" that witnessed the signing of a historic treaty. However, it is a tree that marks an important traditional cultural property for the Comanche.

Site and Tree Threatened

In the late 1990s or early 2000, Linda Pelon took Arterberry to look at the old red oak at the top of the ridge. Because of his extensive experience and knowledge of Comanche culture, Arterberry quickly realized the significance of the site. He recalled hearing about the Storytelling Place, and he was able to connect his people's past presence and history with this sacred site.

The site is a point of high elevation that contains a mixed hardwood forest with prairie grass and wildflowers in the open areas. The elevation changes quickly in some areas from high points to steep hills and bottomland forest in the lower areas. A very diverse mix of plants and trees create a unique ecosystem, which provides habitats and food for a variety of wildlife.

The site contains a round depression or hollowed-out area in white rock, a perfect storytelling place created by nature. The layer of white rock on the surface, coupled with the hollowed-out area, provided a unique site of great significance to the Comanche. A part of their oral tradition involved gathering around a campfire to tell stories of their history, such as tales of great battles, important individuals, or life's lessons to be passed along. The ideal time for storytelling at this site was during a full moon, when the light would have reflected off the white rock, creating a natural lighting for the storytellers.

After a site such as the Storytelling Place has been used consistently over the years, it becomes a part of Comanche traditional ancestral homelands. (The documents that follow further describe the importance and use of the site.)

Office of the Chairman

PROCLAMATION

Special Recognition of a "Comanche Storytelling Place" in Dallas County, Texas

WHEREAS,   The Comanche Nation has existed since time immemorial, long predates the existence of the Nation, establishes the inherent sovereign powers and rights of Comanche self-governance: and

WHEREAS,   The Comanche Nation is dedicated to progress and enhancement of the people of the Comanche Nation: and

WHEREAS,   Dallas County, Texas, contains a "Comanche Storytelling Place", considered a legacy and inheritance of generations to come: and

WHEREAS,   It is the responsibility of the Comanche people to protect this inheritance, like those before, as evidenced in the oral traditions relating to this sacred site: and

WHEREAS,   This "Comanche Storytelling Place", has long been considered part of the Traditional Ancestral homelands of the Comanche people since creation, as related by Comanche elders: and

THEREFORE, I Johnny Wauqua, Chairman of the Comanche Nation, do hereby Proclaim this Comanche Storytelling Place a sacred site and inheritance of the Comanche cultural legacy.

Given Under my hand and
The Great Seal of the
Comanche Nation
On this 23rd day of May, 2002

Johnny Wauqua
Chairman

Comanche Indian Tribe P.O. Box 908 / Lawton, OK 73502

Special recognition of the site elevated its status as a Traditional Cultural Property.

Soon after the important visit with Arterberry to the site, Pelon learned of plans for a future Dallas Area Rapid Transit (DART) line that threatened the loss of the site and the tree. She investigated further to find that unless the site was declared historic or a traditional cultural property, it would not be saved.

Concerned, she contacted Arterberry. On April 18, 2002, he responded by mailing a formal letter on Comanche Nation letterhead to the DART project manager, further describing the importance of the site. The letter was a necessary first step to inform DART of the concerns about the projected rail line. The next step required formal recognition of the site as a traditional cultural property. As a result, Arterberry and Pelon worked to develop the required documentation, and Comanche Nation Chairman Johnny Wauqua provided an official proclamation on May 23, 2002.

On August 6, 2002, Comanche Nation Historic Preservation Officer Jimmy W. Arterberry wrote the following briefing about the site.

On August 12, 2002, there was a meeting that included Arterberry, DART officials, and Pelon, among other concerned individuals. An agreement was reached among the parties, and Arterberry penned a follow-up letter on August 16, 2002. The letter recognized the agreement and specified that minimum impact occur on the sacred site and stipulated that no further advancement be made into the Comanche Storytelling Place. The letter also signified the intent to elevate recognition of the site as a National District Traditional Cultural Property.

Result

As a result of these efforts, the site and the Storytelling Place tree were effectively preserved from the future DART rail line expansion. However, the rail line was not moved far from the tree.

Formal Recognition

Although the site and the red oak were preserved, there was very little known about the history of the tree at the time. As our research progressed, it became clear that the red oak was old enough to qualify as a Comanche

Briefing on A Comanche Storytelling Place
By Jimmy W. Arterberry
August 6, 2002

"A long time ago, it is said" (so-bvay-tsu), begins all Comanche stories.[1] In the oral tradition of creation, migration and shared relationships, Comanche people identify the importance of time, place and social purpose.

Life for a child starts exactly as the ancestors did, by oral instruction and visual aid. Comanche culture relies on the ability to identify the elements of a social structure and the understanding of those relationships.

Storytelling places are conducive to the psychological and physiological training of Comanche children, and considered primary to the overall well-being of man and nature.

Consensus of the Comanche people, to the location of storytelling places, enters into sacred dimensions, which are to be discussed only by the Comanche people. The Trinity Forest storytelling place is located within the bounds of the required elemental characteristics for determining the location of a preferred campsite. Water supply, food sources, shelter, safety and access to forage for horses are required characteristics, for such a site.[2]

The Great Trinity Forests' White Rock Heritage District has been established as a "Comanche Storytelling Place", by proclamation of the Comanche people.[3]

Ethno historical research, archeological documentation, cartographic documentation and consultation between the Comanche Nation and the U.S. Department of the Army, Corps of Engineers, emphatically demonstrates the necessary components for the selected audience, are found at this preferred location.[4]

Shared elemental components are expressed through idioms used in all stories of the Comanche people, emphasizing the importance of relationships between man and nature. The limestone component found at the Great Trinity Forest Storytelling Place, located on the ridgeline in Devon Anderson Park, within the White Rock Heritage District, determines the type and amount of information that is to be presented, and at what time of the day and year that information could and would be made available. There are many traditional ways, in making the determination as to where and when these lessons should and could occur. Relationships between all of the elemental components, the receiving component included, would be defined and refined through the use of storytelling at the chosen location.

The presence of Comanche people in the state of Texas and particularly in the Dallas area, in relation to this writing, has been illusive to the scientific communities in recent times, although descendants of the former inhabitants recognized by the aforementioned in their previous processes, continue to exercise inherent values described in relation to the geography.

Comanche people inherently conceive the importance of honoring the traditional values of the ancestors. Without defined relationships and traditional values, nature will not survive man, man will not survive nature and the results shall be exact, furthering the significance of identifying and protecting such a site, as the Dallas County, Texas, Comanche Storytelling Place, with all of its vital components.

---

[1] Comanche Texts, Elliot Canonge, University of Oklahoma 1958
[2] The Comanches, Lords of the South Plains, Wallace & Hoebel, University of Oklahoma Press, 1952, pg. 14
[3] Comanche Nation Proclamation, 2002
[4] Native American Territorial Ranges in the Central Region of Texas, A Report Prepared to Support NAGPRA Consultation, May 2001, Published by the U.S. Army Corps of Engineers, Ft. Worth District

made available. There are many traditional ways, in making the determination as to where and when these lessons should and could occur. Relationships between all of the elemental components, the receiving component included, would be defined and refined through the use of storytelling at the chosen location.

The presence of Comanche people in the state of Texas and particularly in the Dallas area, in relation to this writing, has been illusive to the scientific communities in recent times, although descendants of the former inhabitants recognized by the aforementioned in their previous processes, continue to exercise inherent values described in relation to the geography.

Comanche people inherently conceive the importance of honoring the traditional values of the ancestors. Without defined relationships and traditional values, nature will not survive man, man will not survive nature and the results shall be exact, furthering the significance of identifying and protecting such a site, as the Dallas County, Texas, Comanche Storytelling Place, with all of its vital components.

The briefing clearly defined the importance of the site to Comanche culture.

Marker Tree. It was also determined that many of the red oaks in the nearby grove were quite old. As a result, the red oak and its associated grove of red oaks were formally recognized by Comanche Nation Chairman Wallace Coffey in a proclamation dated March 5, 2013.

Recognition of the first Comanche Marker Tree without a bend or odd-shaped trunk was a significant event. With the exception of trees related to the previous signing of a treaty, we were not aware that a normal-shaped tree could be a Comanche Marker Tree.

The site would not have been recognized or preserved without the assistance of the Comanche Nation and the cultural knowledge of Jimmy W. Arterberry. At the same time, recognition of the red oak and the associated grove would not have been possible without research regarding the potential age of the trees.

Given the difficulties the Comanche Nation, community leaders, and preservation advocates, including the coalition, encountered in preserving the sacred site and its trees, it seemed especially fitting to honor and respect them with a proclamation recognizing the significance of their history.

A stone retaining wall and fence were moved about 40 yards to avoid the tree.

A tall fence was required due to the drop in elevation to the DART rail below. The tree on the left is not the Storytelling Place tree but another similar red oak.

PROCLAMATION

WHEREAS, the Comanche Indian Tribe is a sovereign nation of free and independent people; and

WHEREAS, the Comanche Indian Tribe, whose ancestors established an independent reputation as the "Lords of the Plains," takes pride in our historic past; and

WHEREAS, The Comanche people occupied and protected the watershed of the Trinity River, called Pih-heet Pah-e-hoona or Three Rivers by our Comanche People; and

WHEREAS, The site of the Storytelling Place in the Great Trinity Forest in Dallas, Texas where this red oak tree grows has been recognized and protected by the Comanche Nation as a sacred Storytelling Place for our Comanche People and is considered a legacy and inheritance of generations to come; and

WHEREAS, It is the responsibility of the Comanche People to continue to protect this inheritance , and the natural features in association with it, as evidenced in the oral traditions related to this site; and

WHEREAS, the Comanche Indian Tribe joins the Dallas Historic Tree Coalition, the City of Dallas, the State of Texas, and the United States government in recognizing the Storytelling Place as important to Native American Cultural Heritage; and

NOW THERFORE, I, WALLACE COFFEY, Chairman of the Comanche Nation, under the authority vested in me by the Constitution of the Comanche Indian Tribe, hereby proclaim and recognize the historic status of Red Oak Grove surrounding the Great Trinity Forest Storytelling Place and proclaim the oldest tree in that Red Oak Grove as:

THE GREAT TRINITY FOREST
COMANCHE STORYTELLING PLACE RED OAK GROVE

In Comanche Country, U.S.A

Given under my hand and seal of the Comanche Nation this 5ᵗʰ day March, 2013

Wallace Coffey, Chairman

Although the Storytelling Place was recognized earlier by the Comanche Nation, the red oak grove and red oak tree lacked formal recognition until 2013.

Cedar Ridge
Comanche Marker Tree

STEVE HOUSER

CHAPTER 11

Arborist Perspective

The Cedar Ridge Comanche Marker Tree is an Eastern red cedar (*Juniperus virginiana*), a species that can attain heights of 70 feet or more and is native to East, North, and South Texas, excluding the Coastal Plains. The species is adapted to acidic and alkaline soils. Individual plants are either male or female. Eastern red cedars with both male and female

Cedar Ridge Comanche Marker Tree.

127

flowers are rare exceptions. The female produces dark blue fruit in the fall and early winter, providing food for birds and other wildlife.

The Eastern red cedar has a pleasing aromatic scent. The heartwood, or straws, in the center portion of the wood is a very attractive color of red with the outer sapwood being white. The aroma reportedly repels insects. The wood is strong and resistant to decay, with the result that early settlers often used the limbs as fence posts.

When Eastern red cedars are grown in a typical forest stand situation, competing with many other trees for light, they often become single-trunk trees. They have a pyramidal, or cone-shaped, growth habit with the highest growth often coming to a point at the top. This growth habit creates a very long leader limb near the top that is often very straight. These qualities made the trees ideal lodge poles for teepees. The tree species was also of great importance in Comanche culture and common rituals.

When our coalition first became aware of the Cedar Ridge Comanche Marker Tree, it was owned by a very kind elderly gentleman who clearly recognized the importance of the tree. Unfortunately, the gentleman is no longer with us, and we do not currently have access to the property.

Comanche Language and Cultural Preservation Committee Members Visit

Eastern red cedar is an important species, and the discovery of a cedar with a sharp bend in the trunk created quite an interest on the part of the Comanche as well as the coalition. Arrangements were made for members of the Comanche Language and Cultural Preservation Committee to visit this site in 2002 and several other places in the Dallas areas thought to be significant to their cultural heritage. The response of these Comanche elders to the Cedar Ridge tree was unusual. As they approached the tree, they spontaneously circled the red cedar and joined hands. As they gathered, they sang to the tree in their language, quantifying its significance to Comanche culture and heritage. A single photo documents the event.

Tree and Site Inspection

Since we do not currently have access to the tree, the following information comes from inspections and observations made in the late 1990s

Comanche Nation representatives visit the Cedar Ridge tree in 2002.

and early 2000. The landowner moved to the property in 1952. He loved the tree and wanted others to know about its existence. Out of respect to his family and the current property owner, his name and the site location remain anonymous.

Close observation and study of the tree provided us with additional education on the growth rate of bent trees in particular circumstances. In this case, the soil is alkaline with shallow rock, and the tree grows on a site that is elevated from its surroundings, with limited water runoff. These factors would have slowed the growth of the tree. There is also a sugarberry (*Celtis laevigata*) touching the base of the red cedar, blocking the sun to the cedar throughout most of the day. The root system of the sugarberry is also in competition with that of the cedar for nutrients and water. (Since sugarberry is a favorite of local birds, it is likely that they brought the sugarberry seeds to the location.) Sugarberries have a moderate growth rate but a shorter lifespan, 50–70 years on average. Given the difficult soil conditions, the estimated age range for this particular sugarberry could be as low as forty-five years, or possibly as high as fifty-five years.

The bent red cedar in the shadow of its sugarberry companion.

The other side of the base exhibits fewer wounds.

There is an old wound at the base of the red cedar that extends from the base along the bend and further out onto the trunk.

The wound at the tree's base is extensive, and it has left shards of the tree protruding from the trunk. This wound undoubtedly caused the loss of many straws, which are required to keep a tree healthy, and slowed the growth rate. The growth was slowed even further by the severe compression and bending of the surviving straws. The vertical lines in the bark are the result of the force created by the bend, or compression scars.

Since the foliage (or needles) produces food for the tree by using light for photosynthesis, the fact that the cedar receives a limited amount of sun would also have slowed the growth in recent years.

Multiple indentions and protrusions are observable along the trunk. The source or cause for them is not clear. They are marked with the photographer's hammer handle and chisel handle in the photo below. The indentions and protrusions could have been created by nature or by wounds created in shaping the tree. Since the trunk and limbs did not touch

The blackened area along the top of the trunk indicates an old wound.

Bark patterns often indicate the past stress a tree has encountered.

the ground as they bowed outward for over 25 feet, the tree was subject through the years to wind moving it in all directions and to ice or snow buildup weighing it down. Trees often react to these external stresses and movements by building up more straws for additional strength. In this case, natural or human-created stresses on the trunk caused a ripple effect in the vascular system.

Tree Size

The diameter of the trunk at its narrowest point between the ground and the bend was slightly more than 16 inches at the time of inspection

*The upright portion of the trunk contained an unusual shape, created
by shade and/or natural forces.*

in April 2000. This measurement was taken several feet off the ground,
so it does not reflect the accurate size of the trunk where it flares greatly
at ground level. The root flare had developed an odd shape from rubbing
against the adjoining sugarberry. At the base, the trunk measured 23 inches
in one direction and slightly over 21 inches in the other direction, or an
average of 22 inches in diameter.

At the time of our measurements, the height of the tree was 26 feet,
and the crown spread (not including the trunk) was an average of 28 feet.
The trunk extended well outside the canopy of the foliage due to the bend.

Multiple indentions followed by protrusions along the trunk.

From the base of the tree to the point at which the trunk starts to bend upward measured 10 feet, and from the base to the outermost reach of the foliage measured 25 feet.

Dating the Age Range

The trunk was long with no real lateral limbs growing off of it until after the upward bend in the trunk. There were two cut wounds in this area that had been left as stumps, which allowed researchers to remove two smaller tree cookies.

Growth rings average 25 rings per inch on one tree cookie sample and 31 rings on the other.

The average growth rate required to reach 144 years old is 13.09 growth rings per inch (144 divided by the radius of 11 inches). Based on current growth ring evidence and the significant wound at the base, the Eastern red cedar was judged to be old enough to qualify as a Comanche Marker Tree.

Eastern red cedars were used by the Comanche for a number of important purposes, but this is the only tree of this species yet submitted as a po-

The cut wounds were around a foot off the ground. The logs in the photo did not relate to this tree.

tential Comanche Marker Tree. The tree also provided some of the tightest growth ring samples obtained from any tree in our research project. It is a unique tree that has taught us a great deal that is of vast importance to our investigations of Eastern red cedars. We also learned that a smaller tree of this species cannot be ruled out without growth ring research and evidence.

Growth rings were quite small but easy to read.

Growth rings were slightly larger on this sample.

California Crossing
Comanche Marker Tree

STEVE HOUSER AND LINDA PELON

CHAPTER 12

Background

BY LINDA PELON

A pecan tree on the grounds of the National Guard Armory adjacent to California Crossing Park was nominated by Irving historian Jim Dunkley following media coverage of the dedication of the Gateway Park Tree in 1997. Dunkley's work on the Bird's Fort Trail has documented the old low-water crossing of the Elm Fork of the Trinity River (called

California Crossing Comanche Marker Tree.

137

California Crossing because it was used by gold seekers traveling to the California Gold Rush of 1849). This trail and crossing were used earlier by Indian guides to take Sam Houston, president of the Republic of Texas, to Bird's Fort for a treaty talk with Texas Indians in 1843.

Dunkley reported to our coalition that an old bent pecan tree similar to the Gateway Park Tree was growing along the old trail, seeming to "point" toward the low-water crossing. Investigations by Steve Houser put the tree in an age range to potentially be old enough to have been used as a marker tree. This information, when combined with Dunkley's research on the history of the trail, resulted in the dedication of this tree as a Comanche Marker Tree by the Dallas Historic Tree Coalition. We were all new to this process, and we didn't think to request a proclamation from the Comanche Nation. We did know, though, that it was a bent pecan tree similar in shape and in age to the Gateway Park pecan and that it grew in a place of historical significance to Texas Indians. There was also documentation of the use of a nearby high point as a Comanche surveillance point and hunting ground. Since the tree's dedication, two groups of Comanche elders, the Comanche Language and Cultural Preservation Committee and the Comanche Nation Tribal Elder Council, have been to visit the tree and offer their blessings.

History

BY STEVE HOUSER

The California Crossing is a historic and well-known crossing point on the Trinity River used by American Indians as well as early settlers. It was also used during the "forty-niner" (1849) gold rush as part of the trail heading west toward California. The trail also led to Turkey Knob, the highest point in the area and a favorite hunting ground for American Indians.

A *Dallas Morning News* article by editorial staff writer Wayne Gard on June 22, 1955, published under the headline "University Campus on Historic Site," notes, "For many Forty-Niners and early emigrants heading west in covered wagons, Turkey Knob was a landmark. The knoll was just about a mile south of the California Crossing of the Elm Fork, used by many of those who pointed their wagon tongues toward Fort Concho and the mesquite and sagebrush country."

A *Dallas Times Herald* article by staff writer Darwin Payne, published on March 22, 1965, under the headline "'Knob' Landmark All But Forgotten," notes:

> But back in 1849 and the surrounding years the famous '49ers relied on Turkey Knob as a landmark to guide them to the popular "California Crossing" on the Elm Fork of the Trinity River.
>
> They knew that just one mile north of the conical hill there was an easy crossing of the river. From there their journey to the promised gold in California could continue via Birdville in Tarrant County and out past Fort Concho at present-day San Angelo.
>
> The river crossing today is remembered in a road there that is still called California Crossing. A park marks the approximate spot of the actual crossing, although a dam has changed the river's nature.

Arborist Perspective

The pecan (*Carya illinoensis*) is the state tree of Texas and also a native species. It can be found from the Piney Woods westward throughout Central Texas and the watershed of the Colorado River. It is a valuable shade tree and important food source today, as it was for the Comanche and other American Indian tribes in the past. It is a tall-growing tree, attaining heights of over 100 feet and a crown spread of over 80 feet. When grown in the open, it forms a large, rounded, symmetrical top. Many different cultivars, or named varieties, exist, each with various sizes and shapes of Pecan nuts, or seeds, which vary in shell size and thickness. Quite a few modern cultivars are named after American Indian tribes—Caddo, Chickasaw, Choctaw, and Cherokee, for example. The California Crossing Tree, as well as all other recognized Comanche Marker Trees that are pecans, are native pecans and pre-date the cultivated varieties.

Tree and Site Inspection

The physical location of the tree is in the 1700 block of California Crossing in Dallas. The tree grows in front of a National Guard building, where a tank often watches over it.

The length of the trunk is 25 feet from the left side, shown in the photo

The tree is located in a grove of pecan intermixed with other species.

above, to the point where it touches the ground and grows upward. The two stems (or limbs) growing upward off the main truck are often referred to as *verticals* or *uprights*. The overall height of the tree is 43 feet and the spread of the crown—the growth from the two limbs plus the adjoining portion of trunk—averages 52 feet. The soil is native black clay, typically moderate to high in alkalinity. There is a very thin layer of organic material on the surface of the soil that averages less than a quarter of an inch.

A significant portion of the base of the trunk is buried under the soil. Here in a floodplain and near the Trinity River, past floods could have left a soil layer over the area as water receded. However, most of the other larger nearby trees have a normal root flare, or root collar, showing at their bases. Because excessive soil buildup from previous floods was not visible on the nearby trees, researchers presumed it to be minimal in the recent past.

The next logical thought about a bent tree in a floodplain is that a past flood may have forced the tree to fall over. When this happens to a larger tree, the collapse typically leaves broken and damaged roots up in the air on the side opposite the fall and often a mound of soil that was pulled from the ground

Normal root flare showing on nearby trees.

as the tree went over. However, there is currently no physical evidence—no upheaved roots or soil—to indicate that this tree fell over in a flood.

In the case of a flood causing a tree to fall over, the tree typically dies or returns to a more upright shape over the years due to the forces of geotropism and phototropism (covered in chapter 8). A mound of soil and roots is always found on the opposite side of a toppled tree; the larger the tree, the larger the uprooted mound. For example, a tree located about 30 yards away from the marker tree shows very clearly that it recently fell over, since the roots and soil are protruding out of the ground.

Tree Size and Theory

Examination has revealed that about 10 feet of the trunk of the marker tree is buried under the soil. Stretching a diameter tape around the section of trunk 10 feet from the base provides a measurement of 26 inches in diameter. The exact diameter at the base of the trunk is unknown, although some logical conclusions can be drawn.

A large mound of soil is created when a tree this size falls over.

Roots are just under the soil surface and are often exposed.

When the California Crossing tree was originally bent, the roots would have been stretched, but not broken to the point of protruding from the soil (as with the comparison tree). This leads to the theory that the California Crossing tree was small when it was bent because there is no upheaval of soil or roots. The trunk diameter was likely less than 10 inches in order for the tree to survive being bent from an upright position to lying on the ground.

Closer Inspection

This particular tree has very little bow to the trunk compared to some marker trees that have long arches. Before the first upright on the trunk, there is an indentation that runs along the trunk on the north side for about 4 feet. Indentations of this nature can be created by external forces (e.g., wind pushing against the trunk). But the indentations could also be related to internal forces, such as the weight of the tree partly resting on the end of the trunk.

The area at the end of a bent or arched trunk—often called the "nose" of the tree—will typically grow upward.

A long indention in the trunk indicates past physical stress.

Two old wounds in the area of a nose.

Since the trunk is bent over, limbs continue to grow outward and often touch the ground on trees with an arched trunk. The photo above shows that there were once limbs growing off the trunk on both the right and left sides, where the trunk touches the ground. It appears that both limbs once touched the ground and, for some reason, died. The most likely causes would be lack of sunlight to these limbs (since they were growing in the shade of limbs above, including the shade of nearby trees) or the weight of the trunk and higher limbs resting on these lower limbs, causing them to crack or break.

Tree Biology

When a tree limb dies, the tree reacts by growing new tissue over the dead limb's base, as shown on the main trunk in the photo above. Each year, the tree adds a small amount of tissue in this area in an effort to wall off and get rid of the dead limb. (Examination of the cut wound about 3 feet above the ground on the trunk in the photo reveals that the tree is growing new tissue around the edges to wall off or "compartmentalize" the wound.) Both limbs once lost at the ground level now are merely protrusions created by the tree's efforts over quite a few years to compartmentalize the wounds. Neither limb appears to have been cut off in the past. It is possible that the trunk once was higher off the ground and rested on one or both of the limbs that have been lost. Once the support of the limbs was lost, the trunk may have fallen over to a greater degree. A tree's root system is designed to support an upright tree, not one bent over to the point of touching the ground. The two limbs on the ground likely acted as crutches to support the trunk, but once they were lost, the tree needed the ground for the additional support the root system could not offer.

As part of a normal inspection, we look closely for abnormal bark lines, scars, and wounds in areas that could have been used to bend the tree. In this case, there is a deep scar at the point from which the first vertical grows upward off the main trunk.

At present, there is not a way to peer inside this wound to determine its

A scar of unknown origin.

The same vertical limb and trunk from another angle. Notice the slight bulge just above the scar where the limb grows upward off the trunk.

origin or how it was created. In studying and researching these trees over time, we have learned that there may be no previous research or documents that explain what we want to know about a tree or a site. We always have many more questions than answers.

With any property we research, we try to "read the landscape" to see the importance of what may have existed long ago. With a tree, it is much the same, in that we try to "read the tree," or see what is not seen by an untrained eye. It may not be clear whether a tree was weighted or tied and retied on a regular basis to hold its form. If a yucca rope was wrapped completely around the trunk, over time it would pinch the straws as the tree expanded in diameter, causing a bulge before the eventual loss of the tree. In this case, the thong was more likely to have been wrapped over the top of the trunk rather than all the way around it.

Consider the difficulty encountered in bending over a small tree to point in a reasonably precise direction using a yucca rope. With no limbs growing off the side of the trunk, it may not have held it in the exact direction intended. An upright or lateral limb growing off a trunk and tied with a yucca rope, would have offered more stability to the trunk as it grew and helped to ensure it was held in the proper direction.

As shown in the previous photo, a scar appears to wrap around the first upward-growing limb. This led us to the theory that any previous rope (or other materials) that created the wound was wrapped not only over the trunk but also stretched further outward beyond the upright limb. It is possible that the material used may not have wrapped over the trunk at all, but around a limb and stretched forward in the direction the tree was being forced to grow.

Although a wound from a yucca rope or any other material would likely have created a scar, once the material was removed or had decayed, the wound would close or compartmentalize. Scars sometimes completely compartmentalize and end up inside the tree and no longer are visible. In other cases, the straws are damaged to the point that a scar or bark disruption persists for many years.

Dating the Age Range

The first upright and second upright growing limbs contained chainsaw or handsaw cut wounds. One of the cut wounds made in the past was approximately 8 inches in diameter, and the other one was 7 inches. The average growth rate near the center of the wounds was 19 rings per inch on one wound and 22 rings per inch on the other. These growth rings were formed during the early stages of their growth, indicating difficult growing conditions. However, more current growth rings on the outer edges of the wounds indicate the growth rate before the limbs were removed improved to around 11 growth rings per inch.

Our researchers removed a small dead limb to help determine a more current growth rate. The site number is always noted on a tree cookie. In this case, the site was number 101.

▶ *Very minor bark disruptions along the trunk that may or may not be of significance.*

▼ *Minor bark disruptions were observed at the base of the second upright limb.*

Smaller growth rings in the center, followed by rings that became larger for unknown reasons.

As decay becomes advanced, growth rings become less evident.

Growth rings in recent years are small, possibly due to recent drought conditions in the Dallas area and much of Texas.

Although the tree cookie was small, it shows a very slow current growth rate for this particular limb at around 20 growth rings per inch.

Since a nearby normal, upright-shaped tree had fallen over in the recent past, there was an opportunity for us to count growth rings on that stump. The species appeared to be a cedar elm *(Ulmus crassifolia)*. Nearby stumps often provide evidence regarding actual and average tree growth at a particular site. The stump was 18 inches in diameter where it was previously cut. It showed a very slow average growth rate of 30 growth rings per inch in the early years, improving to around 11 rings more recently. A growth ring count determined that the recently fallen tree was 112 years old.

One of the difficulties in developing a potential age range for the California Crossing tree being considered as a marker tree is the fact that around 10 feet of the trunk is buried underground. It is not clear how large the actual trunk may be at the base, or when life started for the tree. An assumption must be made regarding the potential taper of the trunk, based on other nearby pecan trees. Assuming another 4 inches in trunk diameter

A fallen tree or stump on a site allows us to collect information on growth rates of nearby trees.

Growth rings are much easier to read on most tree species when live wood tissue is cut.

The growth rings are small in the center but improve in size, which correlates to the growth ring pattern observed on the California Crossing tree.

(or 2 inches in radius) exists in the lower 10 feet of trunk, the pecan would be 30 inches in diameter.

The average growth rate required to reach 144 years old is 9.6 growth rings per inch (144 divided by the radius of 15 inches). Based on current growth ring evidence, the pecan is old enough to qualify as a Comanche Marker Tree.

Direction of Point

The *direction of point* is a term that describes the direction the tree trunk is growing or pointing. A person standing at the base of a tree and holding a compass can get a general idea of the magnetic direction of point by judging the direction of the trunk. The direction of point of this tree was 120 degrees, or east-southeast. Given that the two limbs lying on the ground were likely broken off in the past, the exact angle of the trunk could have shifted slightly when this occurred. The Peters Colony map from 1841 shows the California Crossing and the trail heading to Record Crossing to the southeast.

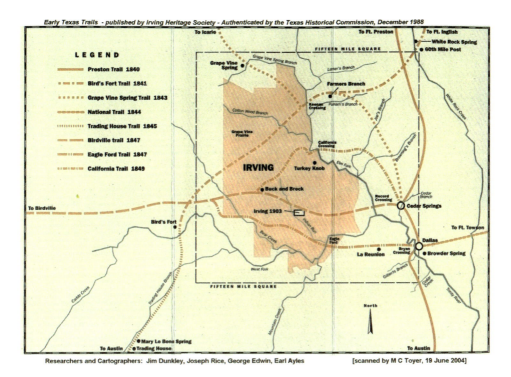

Early trail maps provide clues as to which trees may be related to trails. Courtesy of the Irving Heritage Society.

Comanche Nation Tribal Elder Council Visit

The California Crossing tree, as well as the tree in the following chapter, taught us that trail markers were used to mark a low-water crossing as well as important trails in our area. The nearby tree falling over provided significant growth ring evidence that aided in determining a potential age range. This tree is one of the few that are accessible to the public, and the Comanche Nation Tribal Elder Council members have visited it with enthusiasm.

Side Note

Examination of some of the earlier photos shows areas where bark on the tree has been peeled or removed. When the bark of a pecan is removed, it leaves exposed tissue that is brown to orange in color. Across the street

The coalition was deeply honored to have the Comanche Nation Tribal Council visit the tree in June of 2013.

The Comanche Nation bus, with signage reading "Comanches On The Move."

from the National Guard building is a public park. It is likely that children and possibly even adults were attracted to this tree because it is an easy tree to climb or sit on. Our researchers have several other trees with a similar shape under investigation in public parks. In all cases, bark is missing in places used to climb on the tree. Trees with a low arching bow to the trunk naturally attract children wanting to climb a tree. It seems that parks of the future could contain trees bent in shapes for the purpose of becoming a playground for children.

Bird's Fort Trail Comanche Marker Tree.

Bird's Fort Trail
Comanche Marker Tree

STEVE HOUSER AND LINDA PELON

CHAPTER 13

Background

BY LINDA PELON

A third bent tree, a bur oak on the other side of the low-water crossing of the Elm Fork from the California Crossing Comanche Marker Tree, was nominated for Indian marker tree status in 2012. The tree is located in a previously documented historical and strategic location along the Bird's Fort Trail. More research is needed to gain further understanding of this important area from a Comanche perspective. Like the Lower White Rock Creek area, this area was heavily occupied by Texas Indians, and much of the area is now preserved as parkland that probably contains numerous recorded and unknown archaeological sites. It also fits the template for a preferred Comanche campsite (pecan bottom, freshwater springs, nearby high points). Additional natural features may be identified in the future by knowledgeable tribal leaders as significant to Comanche culture in Texas. This Elm Fork section of the Trinity River corridor may also be eligible as a National Historic Landmark District with Traditional Cultural Properties.

Arborist Perspective

BY STEVE HOUSER

The Bird's Fort Trail Comanche Marker Tree is a bur oak (*Quercus macrocarpa*), a species that can grow to more than 80 feet tall and over 100 feet wide in crown spread. Bur oaks can live for several hundred years, as mentioned in chapter 5. (Information is also found on the Texas Tree Trails

157

Web site.) It is a deciduous tree that produces large leaves and large acorns. It has deeply furrowed bark, and it is native in North Texas as well as in East Texas. It prefers deep rich soils (clay or sand) but adapts well to urban soils.

The soil on this site is black clay and alkaline, with a thin layer of topsoil near the surface (less than 0.25 inches). Researchers observed a minor amount of sand mixed with the topsoil, leading to the conclusion that the area was likely flooded at some point in the past. Flood waters often bring in sand and topsoil from the upper portions of the watershed. Lower sections of the Trinity River floodplain tend to contain higher levels of topsoil, which diminish as elevation increases out of the watershed.

In 2011, David Richardson, a longtime friend of mine and fellow member of the International Oak Society, passed along a photo of the tree. The tree was reported to be in a park in Irving, Texas. Learning more about the specific location, we realized that the tree was just on the other side of the Trinity River and a little north of the California Crossing Comanche Marker Tree.

On our first visit to the park, we found a sign at the parking lot entrance which reads "Bird's Fort Trail." It became clear immediately that the tree was located in the area of the historic trail and that it could potentially be a turning tree for the trail.

Heading west along the existing trail, a traveler would need to cross the river at California Crossing. After crossing the river and heading north, the traveler would find this tree. The tree points to the west, along the trail as it existed at the time in question. (See the Peters Colony map in the California Crossing Tree profile, chapter 12.)

Comanche Nation Tribal Council

In June of 2013, the coalition was deeply honored that the Comanche Nation Tribal Elder Council visited this tree, as well as several other trees.

Tree Size

The overall height of the Bird's Fort Trail tree is 25 feet, and the average crown spread is 43 feet. A significant portion of the trunk is buried under the soil. As the trunk starts to emerge from the ground about 14 feet from its base, the diameter measures 27 inches, and it narrows to 26 inches just

Reconnecting Comanche Nation Elders with their past brings smiles to everyone.

before a knot in the trunk. This leads to the conclusion that the base is much larger by a number of inches. The typical trunk taper for a bur oak of this size, calculated from 14 feet high down to ground level, would add at least 6 inches to the diameter. The actual size could be determined only by carefully excavating the trunk and base, which has led us to some other considerations.

Air Tool

In the future, an air tool could be used carefully to explore what may exist underneath some of the bent trees, such as the Bird's Fort Trail tree. The air tool uses a commercial air compressor to develop high-pressure air that is routed to a wand held by an operator. The air pressure is focused on a concentrated area by the wand, which effectively removes soil while avoiding damage to roots or the trunk. Safety equipment is required during its use as soil is projected in all directions.

Tree Shape

It is important to note that the California Crossing Comanche Marker Tree also has a trunk that is buried under the soil for many feet. Although the shapes are different, given the close proximity of the two trees, it is pos-

An air tool used to excavate a trench near a normal-shaped tree.

Air tool being used to preserve existing roots while trenching.

Root flare exposed with an air tool.

sible that they were bent by the same person or the same Comanche band.

From the point where the trunk of the Bird's Fort Trail tree leaves ground level, the tree has a bow in the trunk and then almost touches the ground again with another bow in the outer limb. This span of trunk adds an additional 15 feet to the trunk's length. With 14 feet being underground and two arching bows, the Bird's Fort Trail tree is the first of this shape to be officially recognized. It is possible that the tree grew more upright at one time in the past. Since the weight of a bent tree is not balanced in the center of the trunk as with a normal tree, it is more likely to have experienced alterations in its shape and form over time.

The limb structure contains two different bows.

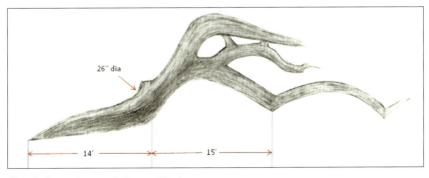

Rough dimensions and shape of the tree.

Direction of Point and Trail

The direction of point is 270 degrees or west, which is the direction of the Bird's Fort Trail depicted in the 1841 Peters Colony map (shown in chapter 14, regarding the California Crossing Tree). Although a number of similar trees around the US are described as *trail trees*, these two should be described as turning trees because each marks a turning point in the trail.

As with the California Crossing tree, this tree does not point in the direction that water flows in the nearby Trinity River. Anytime a tree is near water, our team checks the direction of point for its relationship to the flow of water during a flood. Floodwater can bend small trees, typically the direction that water flows with faster currents. However, most such young trees return to an upright position once the water recedes.

Scars and Wounds

There is a large scar from an old wound on the top of the trunk that runs from the base to 9 feet out on the trunk. The bark patterns indicate the occurrence of multiple wounds along the trunk that coalesced into one long wound.

The source of the long wounds on the trunk is unknown. They have since healed over or compartmentalized. Judging by the bark patterns, the original wound was about 11 inches wide, which would have had a significant impact on the tree's health and, subsequently, its growth rate for quite a number of years. A more recent wound at the base has added to current health concerns, evidenced by several dead limbs in the tree and dieback of foliage in some areas.

Dating the Age Range

Researchers were able to remove two dead limbs and stain them to read the growth rings. The recent growth rate was determined to be 28–30 growth rings per inch.

An old cut wound found on the tree provided further growth ring evidence and a look at the tree's growth rate over a longer period of time. The wound shows very slow growth in its early years (near the center) with a growth rate that averages 24–25 growth rings per inch. The outer edges of

The dark-colored line running down the trunk indicates one of the wounds.

Cross-grain scar at the end of the wound.

Abnormal or wavy bark lines further out on the trunk.

Another cross-grain scar along the trunk.

the wound also show slow growth with an average of 28–30 growth rings per inch. In between the interior and exterior sections, there is a span of healthy growth that averages 13–16 growth rings per inch.

The approximate diameter of the trunk at 14 feet as it emerges from the soil is 27 inches. As previously noted, at least 6 inches in diameter should be added to determine the approximate diameter at the base of the tree,

Tree cookie with tea or coffee added to help read growth rings.

Tree cookie (without any stain) showing a very slow growth rate.

The expanded growth during certain years correlates to the expanded growth rings observed on trees at the nearby California Crossing site.

yielding an approximate diameter of 33 inches and a radius of 16.5. For comparison purposes, the absolute lowest ring count found was 13 rings per inch, which multiplied by the radius of 16.5 equals 214.5 years old.

The average growth rate required to reach 144 years old is 8.73 growth rings per inch (144 divided by the radius of 16.5 inches). Based on current growth ring evidence and the extent of the previous trunk injury, the bur oak is old enough to qualify as a Comanche Marker Tree.

The tree was the first bur oak to be officially recognized, expanding our list of species used for a purpose in the past. This tree also added to our education about turning trees being used to mark turns in trails.

Irving Escarpment Ridge Comanche Marker Tree.

Irving Escarpment Ridge
Comanche Marker Tree

STEVE HOUSER AND LINDA PELON

CHAPTER 14

Background

BY LINDA PELON

An old bent tree was reported on an escarpment ridge that included the Bird's Fort Trail in its viewshed. The first site visit revealed that this tree was growing in a surviving remnant of a historic Texas landscape—the Cross Timbers. This was a scenically beautiful area of red oaks, post oaks, tall grass prairie meadows with wildflowers, and many native plants used as medicines by Comanche healers. The visit to this bent post oak tree revealed that it appeared to mark an adjacent paint rock quarry. This type of rock pigment provided Indians with the color for painting bodies, horses, and hides, and for creating rock art. Researchers found a stone tool in this large quarry, confirming American Indian use of the site. A variety of high-quality pigments (ochre, yellow, orange, and black) were eroding out of the ridge. This quarry is near a strategic high point with documented use by Indians as both a hunting ground for turkey and as a lookout point for observing activities on the trails along the Elm Fork of the Trinity River.

The paint rock quarry and the nearby lookout point/hunting ground are potentially eligible for Traditional Cultural Property status. With permission from the property owners, members of the Comanche Nation Tribal Elder Council visited the site in 2013 and collected some of the pigment rock from the quarry to continue traditional practices with them, including use as artists' materials, cosmetics, and paint for bodies and for horses.

Arborist Perspective

BY STEVE HOUSER

The Irving Escarpment Ridge Comanche Marker Tree was a post oak (*Quercus stellata*), which is a native species in North Central Texas. Post oaks can grow to over 80 feet tall, although on rocky sites they can be much shorter. They have a rounded crown that can spread to more than 70 feet wide. Post oaks are known to be long-lived and slow growing, as well as being drought tolerant. They typically are found on rocky soils along ridges and high points, but they are found in floodplains as well. The acorns are an important food source for wildlife, and the wood was often used as fence posts, hence the name "post" oak.

The area around the tree had been left in a natural state, so native plants and small trees surrounded the Comanche Marker Tree. Our team tries to avoid disturbing any functioning ecosystem as much as possible. As a result, it was very difficult to take a good photograph of the entire tree.

Soil and Native Ecosystem

The soil near the surface is mostly fractured rock that varies in color, with the majority being a shade of red. The soil is shallow (0.25–0.5 inches) due to being on an incline. The soil is slightly acidic and similar to those soils found in the Cross Timbers ecosystem just west of the site and de-scribed in *The Cast Iron Forest: A Natural and Cultural History of the North American Cross Timbers* by Richard V. Francaviglia. A very thin layer of organic material exists on the soil surface.

The tree was a short distance from the historic Turkey Knob (noted in earlier chapters) and sat on one of the highest points in the area. All other trees and plants near the tree's location are small and stunted for their species, given data about optimum growth in better soil conditions. The surrounding vegetation includes prickly pear cactus (*Opuntia* spp.), Eastern red cedar (*Juniperus virginiana*), and cedar elm (*Ulmus crassifolia*).

Direction of Point and Site Inspection

The direction of the tree's point measured southeast at 143 degrees (clockwise from north).

The base of the tree contained a very old wound at ground level, which would not be uncommon to find on a severely bent tree. However, above the old wound, four roots grew out of the trunk and reached into the soil. They had effectively replaced the straws lost when the tree was wounded at the ground level many years ago. These buttress roots are generated in response to the severe lean as a means of support and as replacement for roots that were lost in the past on the side opposite the lean. According to Gerard Claus Mattheck in *Trees: The Mechanical Design*, buttress roots "are found in Europe in spruce, beech, and especially in black poplar at the windward or slopeward side of trees, but rarely seen at the lee side where force flows enter by pressing horizontal roots against the soil, thus relieving the sinker roots of their anchoring function" (1991, 46). A tree with a severe lean or one lying on the soil may generate buttress roots on the opposite side.

Note in the photo below that two of the roots exhibit old wounds and are mostly hollow.

Roots growing outward from the trunk in search of support, water, and other elements.

Another root formed an elbow as it grew into the soil.

The photo above, taken from another angle, shows an additional root coming off the trunk at 4–6 inches off the ground. At first glance, it appeared that soil erosion caused the root to be exposed. However, a second root can be seen growing off the trunk at the normal level in the photo above and in the photos below. The additional buttress root growing from the side of the trunk likely developed in response to the bend in the trunk. Poor (rocky) soil conditions may have been a contributing factor, as well.

Tree Size and Detailed Inspection

The trunk diameter at the base and immediately above the root flare and the outward-growing roots measured 28.75 inches. However, the diameter at ground level was 34 inches. The overall height was measured at 33 feet, and the average crown spread was 46 feet. The trunk grew outward 28 feet before there was an upright or vertical limb.

After a long bow in the trunk and two branches growing upward, the trunk rested on the ground as it branched outward in two primary directions.

No buttress roots are found on the leaning side of the trunk. Photo courtesy of Bill Seaman.

Note the arch in the trunk and the hole from a previous wound in the top of the trunk. Photo courtesy of Bill Seaman.

The tree was not very healthy, and its tip growth rate was less than an inch for many of the recent years. Bud scars on the twigs mark the previous year's growth and aid in determining the current year's growth rate. Twigs in the photo below can be enlarged on a computer to view the bud scars.

Tree Loss Provides an Education

Unfortunately, a period of extended heat and drought in the area during 2011 caused the loss of the tree. Since the tree was old and weak in health, as well as growing in a very rocky location, it was not able to withstand the additional stress of extreme weather.

The loss of the tree provided an opportunity for us to "dissect" or cut it up to find as much information as possible about its history. If the tree had been left untouched, any information it contained inside its vascular system

▲ *The trunk was buried in the ground with two limbs growing outward. Photo courtesy of Bill Seaman.*

▲▶ *Bud scars are more visible on some species than others. Photo courtesy of Bill Seaman.*

▶ *Photo courtesy of Bill Seaman.*

would have been lost as the tree decayed. (As the wood decays, the growth rings become difficult or even impossible to read.)

As a matter of courtesy and respect, we asked Comanche Nation Historic Preservation Officer Jimmy W. Arterberry if there were any cultural concerns about cutting up a dead tree to learn its untold story. The property owners were also contacted to obtain their permission. When they both graciously complied, we scheduled a time to visit the site again.

Bill Seaman, a consulting arborist and Texas Historic Tree Coalition trustee, offered his assistance in the next steps of research. By the time all the visit details were worked out, the tree had decayed to a great degree and wood decay fungi were prolific throughout the tree.

There was a narrow 8-inch opening in the top of the trunk from an old wound of unknown origin. Previous inspections had determined this section of the trunk was hollow, but we hoped to find a solid section near the base.

Two hand-drawn sketches of the tree were created. The first sketch notes the diameter of the trunk and primary limbs, in approximately 24-inch segments, from the base of the trunk outward to the ends of the limbs. The second sketch provides an identifying letter and number for the trunk and

Wood decay organisms are nature's recyclers, turning wood tissue into organic material to benefit soil microorganisms. Photo courtesy of Bill Seaman.

Old wound in the trunk at the top of the bow.

primary limbs. These are based on the limbs' directions of point (indicated by a letter) and the approximate location of a section that was removed (a number). Each section we removed from the site has been marked with a permanent marker that corresponds with the locations noted on the map.

We methodically removed sections of the tree, starting from the ends of the limbs and working back toward the trunk sections and base of the tree. As a result, the numbers for each limb or trunk section began on the ends and increased in size as we worked toward the base of the tree.

We noted two cross-grain scars, located just above the point at which the bowed trunk touches the ground.

As we reached the area where the bow in the trunk touched the ground, we observed an area in the center of the trunk with decay in the heartwood. We removed material from the center of the limb that was similar to black charcoal. As we progressed to the larger trunk sections, it became clear that the center of the trunk was hollow and it contained the same black charcoal material on the inside. This pattern continued all the way to the base of the tree and even into the roots exposed above the ground and emanating from the base of the tree.

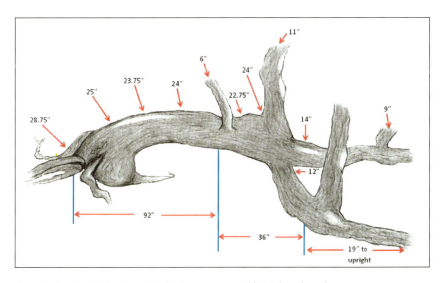

Rough sketch with trunk and limb diameters noted in various locations.

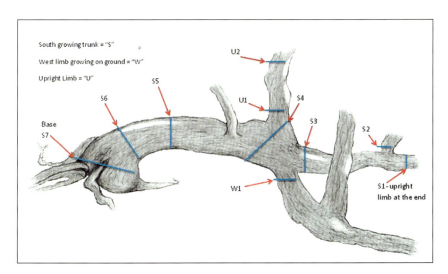

Limb direction noted by a letter—south (S), west (W), or upright (U)—followed by a number to identify the location and section of wood.

Cross-grain scars, which appear to have existed for an extended period of time.

Section of trunk just above the base.

Section removed at ground level showing the hollow base and outward growing roots.

Lightning

Since much of the trunk and many of the above-ground roots were hollow and contained a black charcoal-like material, we suspected that the tree had been struck by lightning at some point in the past. We also considered the possibility of a past fire around or inside the tree, but the evidence did not fully support this theory. Trees seldom survive a fire outside of the trunk and would not likely survive fire placed inside the trunk. It is possible that someone put burning twigs or other burning material inside the crack or old wound on the trunk, presumably to "smoke out" any animal living or staying inside the trunk. Such an occurrence would not have created much fire, due to very limited oxygen inside the trunk, though it would have filled the trunk with smoke. However, the fact that the blackened material extended from the roots to the point at which the trunk touches the ground was not consistent with burning embers or smoke inside the tree. It is most likely that the tree was struck by lightning, since it was located very near the area's highest point. Research shows trees on elevated sites are more likely to be struck. Trees contain a large amount of water within their vascular systems, which makes them good conductors of electricity. Lightning tries to find the quickest way to reach the ground, and a vessel of water (or a tree) offers it an ideal opportunity.

Since the tree had been dead for more than a year, we have considered the possibility that the black charcoal material was related to Hypoxylon Canker (*Hypoxylon atropunctatum*), a secondary pathogen that often affects trees in declining health. The first stages of this infection are light brown or tan in color and dusty to the touch (asexual stage). The infection changes to a silver/gray color, and advanced infections change to a black and hardened material (sexual stage). Further research is required to determine the exact cause of the blackened material.

Bones

As we addressed the base of the tree, we began to find a number of small mammal and bird bones mixed in with humus-like material from decaying organic matter. After we removed the base sections, we collected and photographed the bones.

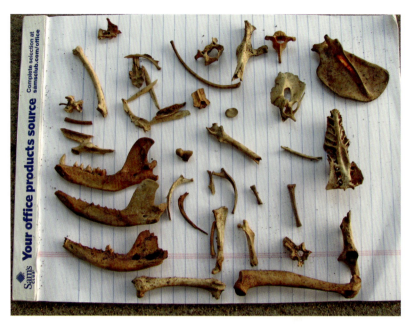

An effort will be made to identify the various wildlife bones.

It is not clear which bones were brought into the tree versus those from wildlife that may have lived in the tree.

Dating the Age Range

Finally, we were able to recover larger limbs that were not hollow and that offered great insight into the tree's history, health, and growth rate.

One of the larger trunk sections without decay in the center.

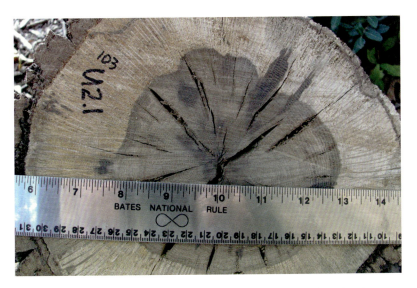

The site number is marked on each section, as well as a letter and number that correspond to the previous hand sketch.

Without any stain or treatment, growth rings are hard to read.

The following are photos of the base section, which was treated with stain to aid in reading the growth rings.

Although the trunk was hollow, we could ascertain that growth rates in the outer portion of the base section ranged from an average of 15–19 growth rings per inch of diameter. Other sections of the trunk contained growth rings as high as 25 per inch.

The average growth rate required to reach 144 years old is 8.47 growth rings per inch (144 divided by the radius of 17 inches). Based on current growth ring evidence and given that the trunk was hollow, the bur oak has been deemed old enough to qualify as a Comanche Marker Tree.

The Irving Escarpment tree added to our education about painted rock quarries, the importance of higher elevations, and the slow growth of hollow-trunked trees. The tragic loss of the tree provided an opportunity to research the growth rings of a bent and hollow tree growing on rocky soil. Each tree we research adds to our education regarding a number and variety of related subjects.

Bottom section with stain consistently shows slow growth.

Bottom section without stain shows growth rings when enlarged on a computer.

Comanche Marker Trees
Conclusion

STEVE HOUSER

The fact remains that additional Indian marker trees still exist today, and there is a need for them to be recognized by the proper authorities. They must be researched and documented before they are lost. There are ways to quantify their potential age without inflicting damage, and there are avenues for uncovering their role in the past.

In researching and recognizing these trees, our base of knowledge on the subject continues to expand. The number of potential Indian marker trees submitted to our coalition continues to grow as word spreads about their existence. Fortunately, the number of volunteers and supporters for the Indian marker tree team also continues to grow. This assistance expands our abilities to research these trees and grows the base of our knowledge, as well as the depth of our understanding. Educating volunteers and the public about Indian marker trees stimulates a consciousness as well as a passion that ultimately drive support for our cause. It is the support of those who understand the importance of our work that makes Indian marker tree research possible.

More information and research will be forthcoming on the subject, as well as on the many interesting trees currently under investigation. Each tree has a history, and some have a truly fascinating story that has yet to be told. If you find a tree in Texas that has potential to qualify as an Indian marker tree, please contact the Texas Historic Tree Coalition (TxHTC.org). Help us to preserve a part of history in your part of the state.

Our goal is for work to continue until all remaining Comanche Marker Trees are found and any imposters ruled out. To honor and respect Comanche Nation heritage—and Texas history as well—is indeed an effort worthy of public support.

Reference List

Downs, Dennis, and Neal Samors. 2011. *Native American Trail Marker Trees: Marking Paths through the Wilderness*. Buffalo Grove, IL: Chicago's Books Press.

Francaviglia, Richard V. 2000. *The Cast Iron Forest: A Natural and Cultural History of the North American Cross Timbers*. Austin: University of Texas Press.

Gelo, Daniel J. 2000. "'Comanche Land and Ever Has Been': A Native Geography of the Nineteenth-Century Comancheria." *Southwestern Historical Quarterly* 103 (3): 273–308.

Gelo, Daniel J., and Wayne J. Pate. 2003. "Texas Indian Trails." Plano, TX: Republic of Texas Press.

Harston, J. Emmor. 1963. *Comanche Land*. San Antonio, TX: Naylor Company.

Jackson, Alvin T. 1938. *Picture-Writing of Texas Indians: Anthropological Papers*. Vol. 2. Austin: University of Texas Publication.

Janssen, Raymond E. 1940. "Trail Signs of the Indians." *Natural History, the Magazine of the American Museum of Natural History*. 45 (January–May).

Jordan, Elaine Blohm. 1997. *Indian Trail Trees*. Ellijay, GA: Jordan Ink Publishing Company.

Larson, Philip R. 1994. *The Vascular Cambium: Development and Structure*. Berlin/Heidelberg and New York: Springer–Verlag.

Marshall, Philip. 2001. "Lopped Trees of Long Island." *Long Island Botanical Society, Quarterly Newsletter* 11 (4): 37, 40–41.

Mattheck, Gerhard Claus. 1991. *Trees: The Mechanical Design*. Berlin/Heidelberg and New York: Springer–Verlag.

McGraw, Joachim A., and Kay Hindes. 1998. *A Texas Legacy: The Old San Antonio Road and Camino Reales, A Tri Centennial History, 1691–1991*. Texas Department of Public Safety.

Mesquite Historical Committee. 1984. *A Stake in the Prairie: The Mesquite Saga; From Frontier Prairie Grasslands to Bustling Space-Age Population Center*. Dallas, TX: Taylor Publishing Company.

Pelon, Linda. 1993. "Issues in Penatuhkah Comanche Ethnohistory." Master's thesis, University of Texas at Arlington.

Schweingruber, F. H. 1993. *Trees and Wood in Dendrochronology*. Berlin Heidelberg: Springer–Verlag.

————. 1998. *Tree Rings: Basics and Applications of Dendrochronology*. Dorecht, Holland: Kluwer Academic Publishers.

Simpson, Benny J. 1999. *A Field Guide to Texas Trees*. Houston, TX: Gulf Publishing.

Stahl, David William. 1990. *The Tree Ring Record of False Spring in Southcentral USA*. University Microfilms International, Arizona State University.

Taylor, Anna Jean, and Melvin Kerchee Sr. (1996). "Searching for Turning Trees and other Special Trees of the Comanche." Unpublished manuscript.

Tekiela, Stan. 2009. *Trees of Texas Field Guide*. Cambridge, MN: Adventure Publications.

Texas Forest Service. 1982. *Famous Trees of Texas*. College Station, TX: Texas Forest Service and Texas A&M University System.

Wells, Don, Diane Wells, John Nardo, Robert Wells, and Lamar Marshall. 2011. *Mystery of the Trees: Native American Makers of a Cultural Way of Life That Soon May Be Gone*. Jasper, GA: Mountain Stewards Publishing Company.

Contributors

JIMMY W. ARTERBERRY

Contemporary American Indian artist Jimmy W. Arterberry is a historian of his tribe and a member of the Antelope band. Arterberry is a direct descendant of preservationists and leaders Niyah, Habby Workee, Sauty Poco, Kiowa Yoko, and Nappywat of the Comanche tribe.

Arterberry is a Comanche scholar and artist, whose experience as the Comanche Nation Tribal Historic Preservation Officer and adjunct professor at the Comanche Nation College (where he teaches Comanche history) provides him the opportunity to address cultural preservation issues, such as sacred site protection, repatriation, and environmental justice.

Arterberry holds a master of arts from the University of Oklahoma, a bachelor of arts from the University of Science and Arts of Oklahoma, and an associate of fine arts as well as an associate of arts from the Institute of American Indian Arts, Santa Fe, New Mexico.

LINDA PELON

Linda Pelon is an anthropologist, ethnohistorian, and educator with a variety of interests and areas of expertise. The early years of her career were focused on assisting individuals with challenges to functioning and successful adaptation. She developed programs in parole rehabilitation to assist recently released ex-offenders with educational enhancement, vocational training, and integration into mainstream culture. She also developed programs and resources to assist individuals coping with closed head injuries and psychiatric illness to achieve enhanced functioning and to accomplish vocational and personal goals. She created an award-winning arts project for individuals coping with psychiatric illness that continues as a consumer-managed project. Her work as executive director for the Dallas Historic Tree Coalition has included research necessary to prepare historic tree nominations. She also has coordinated community advocacy efforts to protect Dallas's urban forest and has served on environmental preservation advisory committees for the city of Dallas.

Pelon's graduate studies research was focused on Texas Comanche history and land use and on assisting the Comanche Nation with reconnecting with their Texas cultural heritage, sacred places, and traditional cultural properties. She continued to pursue these interests and activities as she completed coursework toward a PhD in transatlantic history. Her work with the Comanche Nation continues through teaching and mentoring students. She was recognized in 1997 as an honorary ambassador for the Comanche Nation for her work in recognizing and protecting Comanche Traditional Cultural Properties in Texas. She is currently a professor of anthropology and history at McLennan Community College in Waco, Texas. She also volunteers as the anthropology and history research coordinator for the Texas Historic Tree Coalition. The current focus of this research is to identify Comanche marker trees in Texas and understand Comanche land uses associated with them.

STEVE HOUSER

Life in the tree care business began for Dallas native Steve Houser while he attended morning college classes and simultaneously spent afternoons maintaining the trees at Brookhaven Country Club in Farmers Branch, Texas. Nights and weekends he fed his passion for flying and became a licensed FAA aviation mechanic, attended pilot ground school, and scored 100 percent on the Federal Aviation Administration exam.

One of the first certified arborists in Texas, Houser is owner/president of a Dallas-Fort Worth area tree care firm that he started thirty-four years ago. The firm employs many of the state's top consulting arborists, as well as national and international award-winning tree climbers. He is well regarded by his peers within the industry; the Steve Houser Award of Excellence (aka "The Houser Cup"), established in 2001, is awarded each year to the Texas state tree-climbing champion by the International Society of Arboriculture, Texas chapter.

In his spare time, Houser contributes more than one thousand volunteer hours each year toward worthwhile tree-related causes, educating the public about trees and natural resource-related issues. He teaches forest ecology to master naturalists, and biology and arboriculture to master gardeners in North Central Texas. He has been an advisor for Vision North Texas since its inception and works diligently to encourage responsible and sustainable land use.

Index

Note: Page numbers in *italics* indicate photographs and illustrations.